A Mathematica Primer for Physicists

Textbook Series in Physical Sciences

Understanding Nanomaterials
Malkiat S. Johal

Concise Optics: Concepts, Examples, and Problems
Ajawad I. Haija, M. Z. Numan, W. Larry Freeman

Understanding Nanomaterials, Second Edition
Malkiat S. Johal, Lewis E. Johnson

A Mathematica Primer for Physicists
Jim Napolitano

For more information about this series, please visit:
[www.crcpress.com/Textbook-Series-in-Physical-Sciences/book-series/TPHYSCI]

A Mathematica Primer for Physicists

Jim Napolitano

CRC Press is an imprint of the
Taylor & Francis Group, an informa business

CRC Press
Taylor & Francis Group
6000 Broken Sound Parkway NW, Suite 300
Boca Raton, FL 33487-2742

Printed on acid-free paper
Version Date: 20180223

International Standard Book Number-13: 978-1-138-48656-0 (Hardback)
International Standard Book Number-13: 978-1-138-03509-6 (Paperback)

Library of Congress Cataloging-in-Publication Data

Names: Napolitano, Jim, author.
Title: A Mathematica primer for physicists / Jim Napolitano.
Other titles: Textbook series in physical sciences.
Description: Boca Raton, FL : CRC Press, Taylor & Francis Group, [2018] |
Series: Textbook series in physical sciences | Includes index.
Identifiers: LCCN 2017057101| ISBN 9781138035096 (pbk. : alk. paper) | ISBN
1138035092 (pbk. : alk. paper) | ISBN 9781138486560 (hardback ; alk.
paper) | ISBN 1138486566 (hardback : alk. paper)
Subjects: LCSH: Mathematica (Computer file)--Textbooks. | Physics--Data
processing--Textbooks.
Classification: LCC QC20.7.E4 N36 2018 | DDC 530.0285/536--dc23
LC record available at https://lccn.loc.gov/2017057101

Visit the Taylor & Francis Web site at
http://www.taylorandfrancis.com

and the CRC Press Web site at
http://www.crcpress.com

To Cathy

How to read this book

Look for bits of MATHEMATICA commands that are set off from the text and in a different font, for example

```
Solve[{x + y == 2 a, x - y == 2 b}, {x, y}]
```

(To execute this statement, press *Enter* while holding down the *Shift* key.) I generally follow statements like this with the resulting output. In this case,

```
{{x -> a + b, y -> a - b}}
```

The text around these pieces of input and output is there to help you understand the syntax, and to appreciate how to extend it.

Then, look at the example problems. Every chapter concludes with one or more "Physics Examples" that use the material up to that point to solve typical physics problems.

Many commands have "shorthands" and I use them from time to time, more often towards the end of the book. I've included an appendix to help you resolve them.

Preface

Most textbooks come about because the author taught a course that, at least in her/his opinion, worked out well. This book is no exception.

Everyone agrees that a modern physics curriculum should make use of scientific computing, for extending examples, plotting results, and animating solutions, to name just a few applications. It seems best if computing is woven into the fabric of the courses themselves, rather than taught in a separate course which may or may not connect with the rest of the curriculum. For this to work out, however, the courses need a common framework for their scientific computing.

I tried something in the Summer of 2010, while I was on the physics faculty at Rensselaer Polytechnic Institute. I registered a "computing" course for the Fall, and advertised it informally to students by email. I put a cap of 40 students on the course. It filled up within weeks, so I figured I was onto something.

This was a different kind of course. It was worth only one credit hour, and met once a week for two hours. I used the first 15 minutes or so to show simple examples of a basic computing technique, and then gave the students an exercise they needed to complete and turn in by the next morning. Most students completed the exercise before the end of the two-hour class.

I used different computing platforms at first, but after seeing the student response when I tried MATHEMATICA, I stuck with this program. Both the students and I were impressed with the user interface and documentation, so we were able to quickly learn the program and get into the physics.

This textbook follows the style of that course. Each chapter starts with some simple commands, and applies them to physics problems. Then follows a set of problems that makes use of the material in this, and preceding, chapters.

My course was aimed at the level of sophomore, or above, STEM majors. That is, I assumed my students had already taken a year of calculus and a year of introductory physics. Of course, this book could also be a companion volume for instructors teaching an introductory calculus-based physics course.

This book is *not* styled after a user manual. The documentation for MATHEMATICA is quite complete and, in my opinion, easy to use. Instead, I want to give just enough information for the reader to get started, introducing a limited number of new concepts in what I hope is a clear and narrative manner, especially in the Introduction. The following individual chapters draw on material that precedes them, but do not necessarily need to be followed in order.

Indeed, this book should be useful to professional physicists who would like to come up to speed on MATHEMATICA quickly, and then learn to explore the many other ways they could use the program.

My approach is strictly command-driven. I encourage students to experiment with free-form input, and similarly innovative MATHEMATICA options, but I think it is best to understand the underlying structure before making use of these features.

Many of the figures included here use the default colors of MATHEMATICA output, but they would not reprint well, so I've rendered them in grayscale. (Many thanks to Taylor & Francis for the selected color pages!) The colors can generally be understood from the context. On the other hand, consider it an extra incentive to execute the commands shown and see the output for yourself.

MATHEMATICA has many command shorthands, but I don't often use them in this book, and they are not emphasized. My view is that students will pick them up as they become more proficient, and it is easier to learn the language quickly if you can parse the command structure in a consistent way. Appendix B provides a brief list of the commonly used examples.

I've tried to keep the *Physics Examples* at a level accessible to second year physics students. However, if the attendant mathematics is beyond that level (for example, differential equations or matrix manipulations), then this is difficult. Nevertheless, coupled with suitable instruction or other guidance, students ought to be able to follow them.

Some examples and exercises make use of data files. Data files for all examples and exercises are available through the publisher's website under the Downloads/Updates tab at https://www.crcpress.com/9781138035096.

This book would not have been possible without the help and support of several people. Many thanks to Lou M Han, my editor at Taylor & Francis, for approaching me to write this book and carrying me along all through the process. Thanks also to Kelvin Mischo at Wolfram Research who first encouraged me to try MATHEMATICA for my class, and who helped me with some examples; Andy Dorsett, also at Wolfram, who has helped along the way on so many occasions; and to Jeff Miner at Rensselaer, who made it possible for me to continue using MATHEMATICA with my classes. My Temple University colleague, Tsvetelin Tsankov, read an early draft and made many useful comments, as did the anonymous reviewers to whom Lou sent my original proposal. A special thanks to my friend and colleague, Peter Persans, at Rensselaer, who supported my first ideas to teach this course and, as always, encourages me to find better and better ways to educate my students.

Contents

CHAPTER 1

Introduction

CONTENTS

MATHEMATICA is a very powerful tool. It performs computations that use algebra and calculus for symbolic calculation and data manipulation, and includes special functions and sophisticated numerical algorithms. The program gives you access to physical data, including some that needs to be retrieved from outside sources. The plotting and visualization tools are extensive with a complete set of options that can be made to give you any kind of graph that you'd like.

Learning such a tool can be daunting. Where do you begin? How do you know you aren't starting down a path that has little to do with the kinds of problems you need to solve? When you find examples close to your application, does the approach suit your own intuition? Are such examples even documented in a way that makes it easy to learn the approach?

This Primer is aimed at early undergraduate physics students. The approach takes you quickly to the kinds of problems you need to solve in introductory and foundational curricula in physics or related fields. The approach *does not* emphasize sophisticated or subtly clever manipulations. The goal here is not to make you a MATHEMATICA expert, but rather to get you used to using this very powerful tool for a certain class of applications.

Advanced students and physics professionals can also make use of this book. After getting started with the fundamentals of using MATHEMATICA, you can skip through the chapters, referring back to earlier material as necessary. Some guidance is provided to help you with that.

1.1 GETTING STARTED

The best way to learn a program like MATHEMATICA is to just dive into it, so let's do that.

Start up MATHEMATICA however it's done on your computer. What appears on your screen will depend on the software version, and interface options are expanded in later versions, but for now, all you need to do is to create a new *notebook*. You will enter commands into this notebook and execute them. Commands may be executed one by one, or the whole notebook can be executed sequentially, using the drop-down menu. The notebook and its output can be saved in various ways.

As soon as you type into your notebook, a vertical bracket will appear on the right. This bracket defines a *cell*. A cell may contain commands, output, lines of text for documentation, or other entities.

Now try something very simple. Type the following into your notebook:

```
1+1
```

If you follow this with a carriage-return, you'll see something odd. The command does not execute. Instead, the cell lengthens, vertically, allowing you to type in more commands. That is, you execute a cell, not a single command, although it is certainly possible for a cell to contain only one command.

To execute a cell, type carriage-return while holding down the shift key. This may feel peculiar at first, but you'll get used to it quickly. Do this for "1+1", and MATHEMATICA will return with the output "2".

Now try an *assignment* command. Type and execute

```
a=1
```

and MATHEMATICA returns with an output "1". You've just defined a new quantity a and assigned it the value "1". From this point forward, any time you use a, MATHEMATICA will assume you mean "1", until you redefine a to be something else, or if you remove it from the list of known quantities.

Maybe it seems silly to you for MATHEMATICA to give you an output on such a statement. You can suppress output of any MATHEMATICA line by ending it with a semicolon, that is

```
a=1;
```

makes the same assignment but does not provide the output.

Now put some of this together, and type the following into a single cell:

```
a = 3;
b = 4;
x = a + b
x^2
```

MATHEMATICA returns two output lines, namely "7" and "49".

This is all simple enough. You should take a few minutes, now, to try out some variations on these things. For example, writing "x = a b" multiplies "a" and "b" (but make sure you keep a space between "a" and "b"), and "x = a/ b" divides them.

Now, for "x = a/ b", notice the difference in the output when you use "a=3.;" instead of "a=3;" as written above. In the latter case, you have defined "a" to be a *fixed point integer* instead of a *floating point real number*. This is one of the places where computer nomenclature makes a difference in the way MATHEMATICA works.

This last example, replacing 3 with 3., and getting "0.75" instead of "$\frac{3}{4}$", illustrates why we refer to MATHEMATICA as a *symbolic manipulation* program, rather than a numerical calculator program. As you've seen, though, it is nevertheless easy to have MATHEMATICA do a numerical calculation instead of a symbolic one. We'll come back to these again, from time to time, and see other ways to carry through numerical calculations.

1.2 BUILT-IN OBJECTS

As you have guessed, MATHEMATICA contains lots of built-in functions and values. Execute the following set of commands:

```
Pi
N[Pi]
Sin[Pi/2]
```

You should get the outputs "π", "3.14159", and "1". The object called "Pi" is the way MATHEMATICA refers to the transcendental number π. The functions "N" and "Sin" mean "give me the numerical value of the argument" and "give me the sine of the argument", respectively.

Notice that "Pi", "N", and "Sin" all begin with capital letters. Every built-in MATHEMATICA object begins with a capital letter. (It is for this reason, that when you define your own objects, you should consider starting it with a lowercase letter.) Functions like "N" and "Sin" put their arguments in square brackets, *not* parentheses.

Try something similar with the commands "E", "N[E]", and "Log[E^3]"; MATHEMATICA will return "3" for "Log[E^3]".

Now here's something a little more advanced. Try the following:

```
Log[E^n]
```

Instead of returning with "n", MATHEMATICA says "Log[e^n]", that is, it doesn't seem to do anything. The problem is that n could be a complex number, and so e^n could be complex, and there is not enough to determine the phase of the complex number. That is, for $z = re^{i\theta}$ where r and θ are real, the logarithm would be $\log r + i\theta$.

So, to get the answer you would have expected, you want to expand the expression into real and imaginary parts. MATHEMATICA gives you a function to do that. Try the following:

```
ComplexExpand[Log[E^n]]
```

This should simply return "n".

Next, a different built-in function. Type the following line into a fresh cell:

```
Factor[x^2 - y^2]
```

Notice carefully as you type in "Factor". Each character will be echoed in blue, until you complete the word. This is because MATHEMATICA doesn't recognize what you're typing as a built-in object, until the whole thing is entered. This can be very handy for you as you use MATHEMATICA more and more, especially for spotting typographical errors, or realizing that a variable has already been defined or filled in with a number.

Execute the above command, and get "(x - y) (x + y)". You would get the same result in a slightly more general form, namely

```
w = x^n - y^n;
n = 2; Factor[w]
```

You can then change the value of "n" and re-execute the second line if you'd like to factor, say $x^3 - y^3$. On the other hand, this line will redefine "n" for as long as you keep working, or until you change it again. A neater way to achieve the same goal, without redefining n, is to make use of the *replacement* feature in MATHEMATICA:

```
w = x^n - y^n;
Factor[w /. n -> 2]
```

The line segment "/. n->2" means[1] "In what comes before in this command, replace all occurrences of n with 2."

1.3 FUNCTIONS

The built-in MATHEMATICA objects "Sin", "N", "Log", to name a few of those we have used so far, are called *functions*. They are, literally, functions of the arguments enclosed in the square brackets that follow. You can in fact define your own functions in MATHEMATICA. For most beginner applications, though, you won't need them, because expressions will do just fine. Furthermore, the syntax of functions can be tricky, so I don't recommend using functions unless it is clearly the best way to solve your problem. Perhaps, for example, you will find it handy to define a set of different expressions from one or more custom functions.

This is how to define a function:

[1] I wrote in the Preface that this book avoids MATHEMATICA shorthands, but in fact, the "/." construction *is* a shorthand for the command "Replace". For simple replacements, however, the standard is to just use "/.", but it is worth your time to look through the documentation on "Replace" and the transformation rules.

```
f[var_] = var^2;
```

In the normal language of calculus, this defines the function $f(x) = x^2$. Notice that I defined a function “f” with a lowercase letter; this avoids potential conflicts with built-in MATHEMATICA functions. The underline after “var” is called a *blank*. Once this is defined, you can, for example, define an expression $y = x^2$ with

```
y = f[x]
```

or an expression $z = x^3$ with

```
z = f[x^(3/2)]
```

The power behind the use of functions should becoming clear to you. With great power comes great responsibility, though, so be careful how you use functions.

1.4 SIMPLE PLOTTING

It is easy to make simple plots. Executing the commands

```
amp = A;
osc = Sin[Pi x];
func = amp osc
Plot[func /. A -> 10, {x, 0, 10}]
```

outputs the symbolic expression “$A\sin(\pi x)$” and the plot

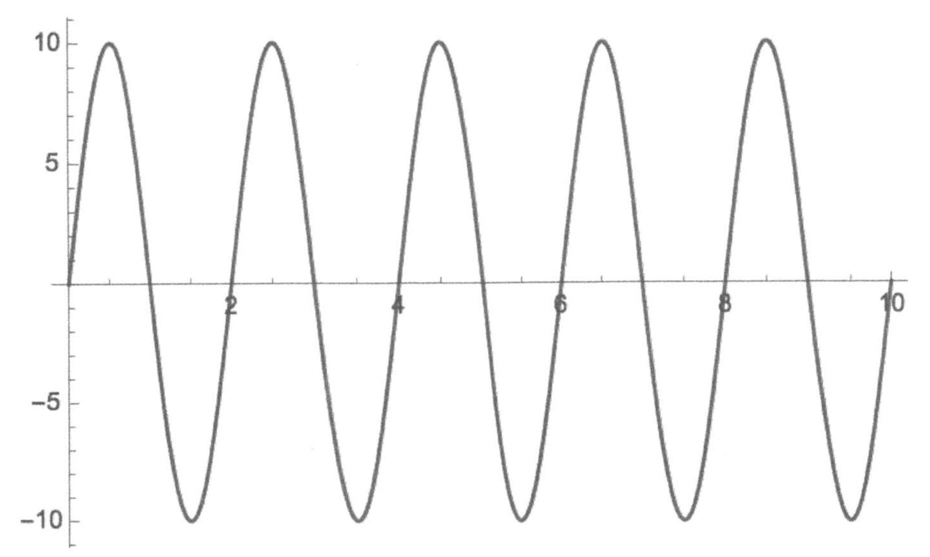

The syntax of “Plot” is pretty much self explanatory, but there are a very large number of options. These include axis scaling; line thickness, color, and

texture; filling options; and axis labeling and style, just to name a few. You can refer to the MATHEMATICA documentation (described later in this chapter) for details.

The expression "{x, 0, 10}" inside "Plot" is your first example of a *list*. Lists are widely used objects in MATHEMATICA, and you will become familiar with them in many contexts. This one clearly specifies the dependent variable, and its lower and upper limits for the plot. A list with these specifications is expected by the "Plot" command. Section 2.2 covers lists in more detail.

Here's a slightly more advanced use of "Plot":

```
amp = A;
tau = a;
osc = Sin[Pi x];
damp = Exp[-x/tau];
func1 = amp damp osc
func2 = amp damp;
func3 = -amp damp;
Plot[{func1, func2, func3} /. {A -> 10, a -> 5}, {x, 0, 10}]
```

This produces the plot

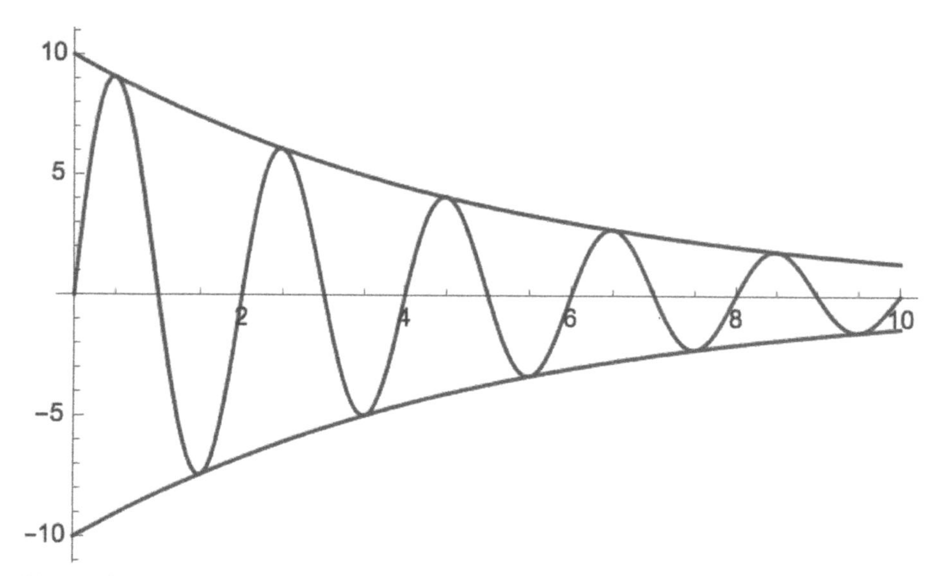

Notice how this example makes more use of lists. The object "{func1, func2, func3}" gives a list of functions to plot, rather than just one. The replacements "{A -> 10, a -> 5}" are also a list.

Physics problems often involve plots that are not simply functions of one (or more) variables, but instead are plots that trace out paths. Common examples are trajectories or orbits, that is, paths that give the position of something as a function of time or some other variable. In these cases, you can use the function "ParametricPlot". Consider the example

```
x = a Cos[q];
y = b Sin[q];
ellipse1 = {x, y} /. {a -> 3, b -> 2};
ellipse2 = {x, y} /. {a -> 3, b -> 1};
ParametricPlot[{ellipse1, ellipse2}, {q, 0, 2 Pi}]
```

Rather obviously, the two expressions "ellipse1" and "ellipse2" represent loci of points "x,y" that are ellipses with semimajor axis $a = 3$ snd semiminor axes $b = 2$ and $b = 1$, defined by values of the parameter q. (If you are wondering why I used q instead of θ, for example, wait until we do the physics example at the end of this chapter.) This cell of commands produces the plot

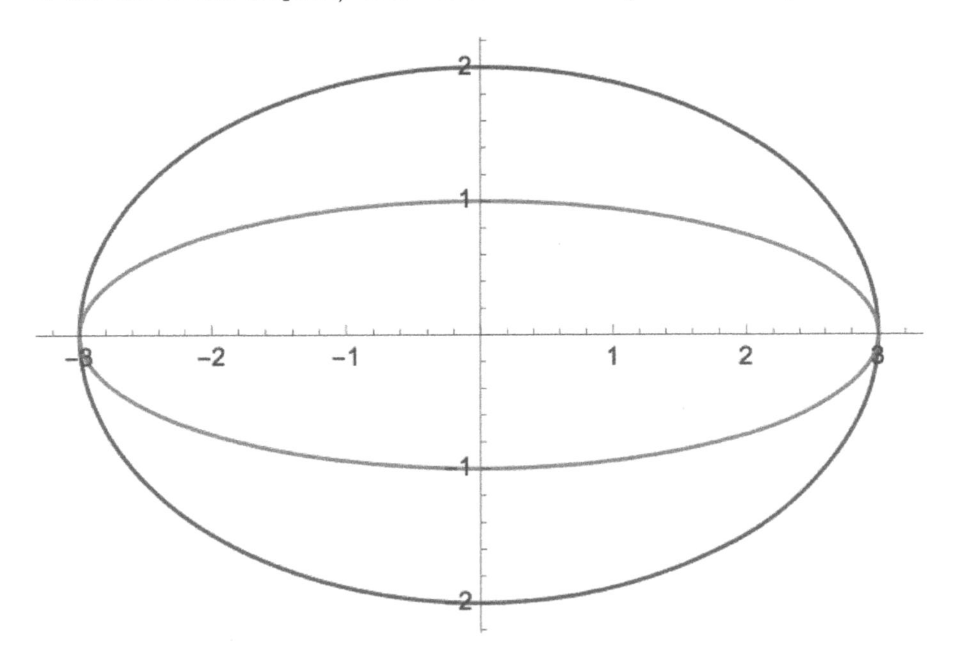

1.5 GOOD HABITS FOR WRITING NOTEBOOKS

You are now ready to start writing your own notebooks for solving physics problems. Before you start in earnest, it's a good idea to learn some good habits.

First, be neat. Arrange your commands in a readable form, and use carriage returns to split commands over lines if necessary. MATHEMATICA will indent things so that they are easily understood.

Most commands in MATHEMATICA end up creating new quantities, and MATHEMATICA remembers these throughout a session. When you write a notebook, it is natural to think of each notebook execution as a fresh start, but quantities left over from the last execution (or anything else you were doing on the side) will remain, and this can lead to confusing results.

For this reason, I recommend starting every notebook with the command

```
Remove["Global`*"]
```

This will clear away all variables defined during a MATHEMATICA session. There are other, less drastic, ways of cleaning up, but, for now, this will do.

Document your notebooks well! You can use cells to contain only text, and the Format→Style menu lets you choose from a group of heading and text styles. There are plenty of other documentation tricks, but you'll learn them as you go along.

1.6 PHYSICS EXAMPLE

Example 1.1 *A resistor R and capacitor C are arranged in series with a switch and a source of EMF V. The capacitor is initially uncharged when the switch is closed, so the charge Q on the capacitor, as a function of time t, is*

$$Q(t) = CV\left(1 - e^{-t/\tau}\right) \tag{1.1}$$

where $\tau = RC$. For $V = 1$ Volt and $C = 1\ \mu F$, plot $Q(t)$ for $t = 0$ to $t = 5$ ms for each of the three values $R = 1\ k\Omega$, $R = 2\ k\Omega$, and $R = 5\ k\Omega$.

A notebook which solves this problem is shown in Notebook 1.1. I am including here the output of the *executed* notebook. In addition to what we've already discussed in this chapter, it makes use of some options to embellish the plot. Let's go through the whole notebook from top to bottom.

Note that you do not need to evaluate each line one by one. It is possible to execute the entire notebook. Choose "Evaluate Notebook" from the "Evaluation" menu, and the notebook is executed from start to finish. There are other options for evaluating cells or sections of a notebook.

The very first line uses the "Remove" command to clear up any previously defined variables, so we start from a clean slate. The next cell uses the "Format/Style" called "Section" to create a title for the notebook. The following cell (and two more below it) use "Subsection" to logically mark off different portions of the notebook.

Just as the heading says, the first subsection defines the quantity we want to plot, and lets MATHEMATICA print out the symbolic form. Notice that it is called "q" while the text called it Q. Similarly for C and V. I am being safe here, taking care to use only lowercase first-characters when defining new variables. Indeed, if you try to use "C" you'll find that MATHEMATICA tells you it is a protected name.

You notice that I managed to put a Greek letter, τ, into the equation. The simplest way to do this is to use the LaTeX code for such symbols, sandwiched in between two "escape" keystrokes. That is, you type

```
<escape>-\tau-<escape>
```

NOTEBOOK 1.1 Solution to Example 1.1

```
In[1]:= Remove["Global`*"]
```

Charging a capacitor

Define the equation

```
In[2]:= τ = r c;
        q = c v (1 - Exp[-t / τ])
```

Out[3]= $c\left(1 - e^{-\frac{t}{c\, r}}\right) v$

Form the three expressions

```
In[4]:= c = 10^(-6); v = 1;
        q1 = q /. r → 1 × 10^3;
        q2 = q /. r → 2 × 10^3;
        q3 = q /. r → 5 × 10^3;
```

Plot the three expressions with scaling

```
In[8]:= qVals = 10^6 {q1, q2, q3} /. t → 10^(-3) tms;
        Plot[qVals, {tms, 0, 5},
         PlotStyle → {Black, {Red, Dashed}, {Blue, DotDashed}},
         PlotLegends → {"1kΩ", "2kΩ", "5kΩ"},
         AxesLabel → {"t (ms)", "Q (μC)"}]
```

Out[9]=

Q (μC)
1.0
0.8
0.6
0.4
0.2
1
2
3
4
5
t (ms)
1kΩ
2kΩ
5kΩ

and the Greek letter τ appears. You'll see that this is used further down in the notebook for the symbols Ω and μ. Another way to enter Greek letters is with the

```
\[...]
```

construction. For example

```
\[Mu]
```

and

```
\[CapitalOmega]
```

appear as the Greek letters μ and Ω.

The next subsection puts in the numbers, all with semicolons terminating the lines so that these simple manipulations are not echoed. Notice that I used the replacement feature to define the three different expressions for Q that we are asked to plot. We stick with SI units, namely Farads, Volts, and Ohms. Expressions like "1×10^3" are typed in without the ×, and MATHEMATICA automatically fills it in.

The last subsection creates the plot. A list "qVals" is created, containing each of the three expressions for $Q(t)$. Since it makes more sense to plot charge in μC instead of Coulombs, and time in milliseconds instead of seconds, the expressions are all multiplied by 10^6 and the time variable "t" is replaced by a variable "tms" multiplied by 10^{-3}. (If you would like to see more explicitly what's happening here, remove the semicolons at the end of the lines.)

The first line of the "Plot" command is what we have already discussed, but the following three lines are new, three options that are available with "Plot". They should be self explanatory, but notice that each of them is a list replacement structure. Also notice that each of the three option lines are indented neatly. (They are entered with carriage returns at the ends of the lines.) Of course, there are many other options, so now is a good time for you to learn how to find out what they are.

I should draw your attention to the "PlotStyle" option specification. It is a list of three elements, one for each of the plots. The second and third list elements, however, are themselves two-element lists which specify the line texture in addition to the color. You can't see the color clearly on this printed page, but the texture is evident.

1.7 GETTING HELP

There is plenty of documentation on MATHEMATICA, both online and within the program itself. There are also user groups to which you can subscribe, to look for solutions and answer questions, and to find relevant examples. Eventually, you'll settle on the path that works best for you.

My recommendation is to start with the help listings that come with the program. When you start up MATHEMATICA, you will see (at least, in MATHEMATICA 10), a "Wolfram MATHEMATICA" window with the "Welcome Screen". There are three buttons across the bottom. The help listings I recommend are the button called "Documentation". This brings you to the "Documentation Center" which is searchable, and also organized by topic. Try it out some time soon, but looking for some things you already know, such as "Plot".

The second button brings you to the "Wolfram Community" list of discussion groups, and the "Resources" button takes you to the many other options for getting help.

You can use the "Help" menu in MATHEMATICA to bring up the "Documentation" listing, or get you back to the Welcome Screen.

One other useful online resource is the Mathematica Stack Exchange:

http://mathematica.stackexchange.com

This website allows you to exchange questions and answers with the worldwide MATHEMATICA community. You'll notice that the answers to many questions are already available to you at that site.

1.8 CHAPTER SUMMARY

- Open a *notebook* to enter statements in MATHEMATICA. To execute a statement, hit *return* while holding down the *shift* key.
- Assignment statements associate some variable with a symbolic or numerical expression.
- Remember that all built-in MATHEMATICA functions start with a capital letter, and that arguments are enclosed in square brackets.
- You can define your own functions, but be careful of the syntax.
- Make a simple plot with "Plot[x^2, {x,0,1}]" and go from there.
- Develop good habits early for writing and documenting notebooks.
- Start using the MATHEMATICA documentation to find your way around and learn more about these techniques.

EXERCISES

1.1 *A projectile is fired with initial speed v_0 and at an angle θ with respect to the horizontal. Neglecting air resistance, its range is given by*

$$R = \frac{2v_0^2}{g} \cos\theta \sin\theta$$

where g is the acceleration due to gravity near the Earth's surface. Make a plot of the range as a function of angle for $v_0 = 30$ m/sec.

1.2 *A projectile is launched with initial speed* v_0 *and at an angle* θ *with respect to the horizontal. Its position* (x, y) *at any time* t *is given by*

$$x(t) = v_0 t \cos\theta \qquad \text{and} \qquad y(t) = v_0 t \sin\theta - gt^2/2$$

where x (y) *measures the horizontal (vertical) position and* g *is the acceleration due to gravity near the Earth's surface. Plot the trajectory, that is, the path in the* (x, y) *plane, for* $v_0 = 30$ *m/sec and* $\theta = 30°$. *Note that it is simple to specify the argument of a trigonometric function in degrees instead of the default radians; check the* MATHEMATICA *documentation to see how.*

1.3 *A simple harmonic oscillator consists of a mass* m *and a spring with stiffness constant* k. *It moves in one dimension* $x(t)$ *with velocity* $v(t)$ *where*

$$\begin{aligned} x(t) &= A\cos(\omega t + \phi) \\ \text{and} \qquad v(t) = \dot{x}(t) &= -\omega A \sin(\omega t + \phi) \end{aligned}$$

with $\omega = \sqrt{k/m}$. *Set* $m = 1.75$ *kg,* $k = 2$ *N/m and make two plots, one for* $x(t)$ *and one for* $v(t)$, *where* $0 \le t \le 10$ *sec. Assume* $A = 5$ *m and* $\phi = \pi/4$.

1.4 *Referring to Exercise 1.3, at the initial time* $t = 0$, *the position and velocity are given by* $x_0 = x(0) = A\cos\phi$ *and* $v_0 = v(0) = -\omega A \sin\phi$ *so that*

$$A = \sqrt{x_0^2 + \frac{v_0^2}{\omega^2}} \qquad \text{and} \qquad \phi = \tan^{-1}\left(-\frac{v_0}{\omega x_0}\right)$$

Make the same plots, but now for an oscillator that starts from $x_0 = 0.25$ *m with velocity* $v_0 = 1.25$ *m/sec. (You should check the* MATHEMATICA *documentation to learn about the function "*ArcTan*".)*

1.5 *The kinetic energy of a mass* m *with velocity* v *is* $K = \frac{1}{2}mv^2$. *For a spring with extension* x *and stiffness* k, *the potential energy is* $U = \frac{1}{2}kx^2$. *Referring to Exercise 1.4, plot on the same graph (a) the kinetic energy of the mass, (b) the potential energy of the spring, and (c) the sum of the two.*

1.6 *The figure below shows an arrangement of three charges:*

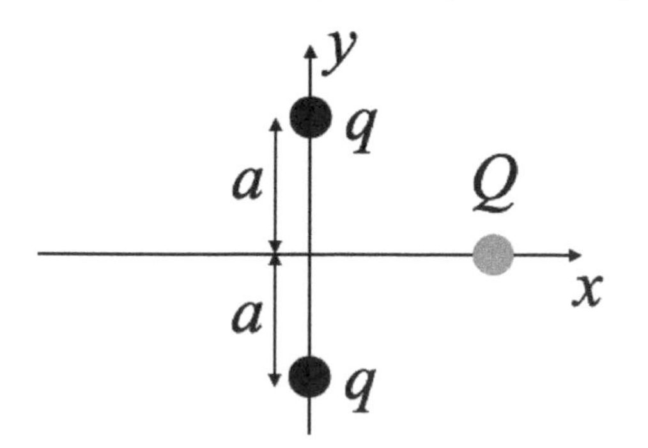

The two charges q are fixed at $y = \pm a$ on the y-axis, while the charge Q is free to move along the x-axis. Find the net force $F(x)$ on Q as a function of x, and make a plot of $F(x)$ for the two cases where Q is the same or opposite sign of q.

1.7 *Planetary orbits are described in plane polar coordinates (r, ϕ) as*

$$r(\phi) = \frac{r_0}{1 + \epsilon \cos \phi}$$

where r_0 is a scale parameter and ϵ is the eccentricity of the orbit. The orbit is a circle for $\epsilon = 0$, an ellipse for $0 < \epsilon < 1$, a parabola for $\epsilon = 1$, and a hyperbola for $\epsilon > 1$. Define a function $r(\phi)$ and expressions $x = x(\phi) = r \cos \phi$ and $y = y(\phi) = r \sin \phi$. Then, choose one value for r_0, and appropriate choices for ϵ, to plot the four orbit shapes on a single graph, using ParametricPlot. *For the hyperbola, plot only the one branch that extends to the same side of the focus (i.e. $r = 0$) as the parabola. (You may need to learn about the function "*Show*" in order to put all the plots on one graph.)*

1.8 *Repeat Exercise 1.7, but this time use the documentation to learn about "*PolarPlot*" and use this to plot the orbits as functions $r(\phi)$ instead of converting to Cartesian coordinates.*

CHAPTER 2

Solving Algebraic Equations

CONTENTS

Equation solving is everywhere in physics. The fact that two expressions equal each other will always imply some other relationship between the elements of that expression. This new relationship in turn implies some physical consequence, and next steps go on from there.

For example, the velocity v of an object thrown straight upward with initial velocity v_0 is just $v = v_0 - gt$. Therefore the time at which the object stops and falls back downward is $t = v_0/g$. Physicists will typically solve such equations without thinking that's what we are doing, but indeed, we arrive at this answer by solving the equation $0 = v_0 - gt$ for t. Of course, the equations we need to solve are often much more complicated, can involve more than one unknown, and might have more than one solution.

The emphasis of this chapter is on the built-in MATHEMATICA function "Solve", which finds analytic solutions to algebraic equations. (Numerical solutions and differential equation solving are the subjects of later chapters.) The techniques are straightforward to learn and to use.

This chapter also slips in some MATHEMATICA concepts, most importantly the use of *lists*. These are just lists of items separated by commas and bounded by the curly brackets { and }. Understanding lists comes up naturally when we try to understand the syntax of the answer returned by "Solve". We'll also take this opportunity to first introduce you to logical expressions (but with more on this in a later chapter) and also the use of complex numbers in MATHEMATICA.

If you look in the Documentation Center under "Equation Solving" (under Mathematics and Algorithms), you'll find more detail and extensions on all of this material.

2.1 SYNTAX FOR EQUATIONS AND SOLUTIONS

Probably the most difficult thing to overcome when learning symbolic manipulation programs like MATHEMATICA, is the syntax for the input and output after you get past the most basic concepts.

Let's try solving a very simple equation to see how things work. Type the following into MATHEMATICA:

```
solution = Solve[x + a == b, x]
```

The first argument of "Solve" is the equation you want to solve, and the second argument is what you want to solve for. Note that the equation is written with a double equals sign, "==". A single equals sign is what we have been using for expression assignment. The double equals sign has a much different meaning; it is the way you write an equation as a logical expression. Once again, even though this may seem a little weird to you, you'll get used to it quickly.

So, the statement above says "Solve the equation $x+a=b$ for the variable x and put the result in the expression called 'solution'." Execute this statement (as before, with a "shift, carriage-return") and you should see the following:

```
{{x -> -a + b}}
```

The solution is a replacement statement stored in an embedded list structure. This is the entity stored in the variable we called "solution". You can prove that to yourself by typing "solution" by itself into a cell and executing it.

The solution to $x+a=b$ is, of course, $x=b-a$, so MATHEMATICA clearly gives the right answer, but why does it give it to you in the form "x → -a + b"? Let's try to see the answer to this question in order, namely "Why a replacement?", "Why as a single list entry?", and "Why as a list embedded in a list?"

If you think about it for a moment, a replacement statement is just what you want. If MATHEMATICA gave you an equation $x=-a+b$, then you'd have the answer, but you couldn't do anything with it. Of course, MATHEMATICA could just make the assignment "x=-a+b", but then you will have lost the variable "x". Giving you a replacement statement makes it easy for you to put in this expression for "x" wherever you need it, for example

```
solution = Solve[x + a == b, x];
x^2 /. solution
```

which returns "{(-a + b)^2}".

Now, sometimes there is more than one solution to an equation. This is why the answer comes as a list. Execute the following

```
Solve[x^2 == a^2, x]
```

and MATHEMATICA returns

```
{{x -> -a}, {x -> a}}
```

Similarly, if you execute

```
Solve[x^3 - 6 x^2 + 11 x - 6 == 0, x]
```

then MATHEMATICA returns

```
{{x -> 1}, {x -> 2}, {x -> 3}}
```

That is, the inner list is a list of the possible solutions to the equation.

Before we try to understand the meaning of the outer list, let's take a little detour and try to better understand lists themselves.

2.2 LIST MANIPULATIONS

This is a good time to take a short detour and talk about *lists* in MATHEMATICA. If you search for "Lists" in the Documentation Center, you'll find plenty of description and examples, so I will keep this discussion brief.

A MATHEMATICA *list* is simply a list of objects, separated by commas and enclosed in braces. There are almost no restrictions on what kind of object can be in a list, and the meaning of these objects depends on the context. Indeed, as we saw in the results of Solve, an object in a list can itself be another list.

Probably the simplest application of a list is to do something to a collection of objects that you might do to any one of them. For example, if you execute

```
{x^2, x^3, x^4}^2
```

then MATHEMATICA returns

```
{x^4, x^6, x^8}
```

Or, if you execute

```
N[Cos[{Pi, Pi/2, Pi/3}]]
```

then MATHEMATICA returns

```
{-1., 0., 0.5}
```

MATHEMATICA provides a number of built-in functions that return *pieces* of lists. They are easily enough found using the program's documentation. Probably the most often used function is "Part", which returns a list element or a group of list elements. For example if you execute the cell

```
list1 = {a, b, c, d};
Part[list1, 2]
```

then MATHEMATICA simply returns "b". That is, the first argument of "Part" is the list, and the second argument tells which element of the list is to be extracted. There are many alternative syntaxes for "Part" for extracting several elements or ranges of elements.

A very common shorthand for "Part" is double square brackets. That is, "Part[list,i]" is equivalent to "list[[i]]". In other words, executing the cell

```
list1 = {a, b, c, d};
list1[[2]]
```

returns the same result as the above example.

We will encounter lists throughout this book, introducing new contexts for them along the way. If you are impatient, you can go right now to the MATHEMATICA tutorial on "Lists" and learn lots more about them.

2.3 SYSTEMS OF EQUATIONS

Now let's get back to solving algebraic equations. The examples so far were about solving one equation for one variable, although sometimes there was more than one solution for the variable.

MATHEMATICA can also solve systems of equations. You can probably guess the syntax by now, with both the system of equations and solution variables contained in lists. Here's a simple example to execute:

```
Solve[{x + y == 2 a, x - y == 2 b}, {x, y}]
```

then MATHEMATICA returns

```
{{x -> a + b, y -> a - b}}
```

This time the inner list contains the solutions for the two solution variables. If you wanted the solutions for x and y separately, you can use "Part" to separate them, but remember that you have nested lists here. For example, if you execute

```
sol = Solve[{x + y == 2 a, x - y == 2 b}, {x, y}];
Part[sol, 1]
```

then you get

```
{x -> a + b, y -> a - b}
```

which is probably not what you wanted. However,

```
Part[Part[sol, 1], 2]
```

returns just the solution for y, that is

```
y -> a - b
```

Similarly "Part[Part[sol, 1], 1]" returns the solution for x.

You should also convince yourself that "sol[[1]][[2]]" is equivalent to "Part[Part[sol, 1], 2]". This alternate form of "Part" can be confusing until you are used to parsing out MATHEMATICA commands, so I recommend sticking with the long form until you are comfortable with the syntax.

This is a good point at which to bring up *logical expressions*. An equation is one example of a logical expression. The "==" sign is a logical operator that means "the expression on the left is the same as the expression on the right." The equation makes a logical statement, and "Solve" is designed to extract a conclusion from that statement.

Another logical operator is "&&" which means "and". That is, it can be sandwiched in between two logical statements forming a new statement which means that both statements are true. This means that an alternate way to write the above use of "Solve" for a system of equations is the following:

```
Solve[x + y == 2 a && x - y == 2 b, {x, y}]
```

which also returns

```
{{x -> a + b, y -> a - b}}
```

Note that the system of equations is now a single logical expression, not contained in a list.

Of course, when it comes to logical expressions, this is just the tip of the iceberg. We'll encounter more of these in this book, but you might also want to explore them using the MATHEMATICA documentation.

We haven't yet gotten around to explaining why "Solve" returns nested lists in the first place, but maybe the reason is obvious to you by this time. We've seen the inner list represent different answers for the same variable, and we've also seen it represent the answers for the different variables in multivariable equations. So, we'll need the nested lists for multivariable equations where there is more than one solution for one or more of the variables.

Here's a very simple example to illustrate this syntax. Execute

```
Solve[{x^2 == a^2, y == x - b}, {x, y}]
```

and MATHEMATICA returns

```
{{x -> -a, y -> -a - b}, {x -> a, y -> a - b}}
```

and you see that the inner lists are for the different variables, while the outer list has the different possible solutions.

2.4 COMPLEX NUMBERS

Before moving on to the physics examples, let's take a quick look at how complex numbers are treated in MATHEMATICA. Execute

```
Solve[x^2 == -1, x]
```

and you get[1]

```
{{x -> -I}, {x -> I}}
```

[1]In fact, MATHEMATICA has a special form for the output form of $i \equiv \sqrt{-1}$, which is neither the variable "i" nor the internal value "I", but it does not translate well into LaTeX.

You see that, in keeping with the notion that capitalized variables are often internally stored values, "I" is the value $\sqrt{-1}$.

Complex numbers are built in the obvious way, for example

```
z = x + I y
```

defines the complex variable $z = x + iy$.

Operations on complex numbers are pretty much what you'd expect, but don't forget that, in principle, x and y can be complex numbers as well. For example,

```
Re[z]
```

returns

```
-Im[y] + Re[x]
```

which of course is technically correct. Probably the easiest way to reduce this general case is to use "ComplexExpand" which expands a complex number assuming that all variables are real. That is

```
ComplexExpand[Re[z]]
```

returns simply "x". Similarly, you find Euler's formula with

```
ComplexExpand[Exp[I t]]
```

which returns "Cos[t] + I Sin[t]". You might recall that we already encountered "Complex Expand" on Page 4.

A more formal way to deal with real and imaginary parts of complex numbers, is to couple the use of "Simplify" with "Assumptions" using the concept of *Domains*. For example

```
Simplify[Im[z],
   Assumptions -> {x \[Element] Reals, y \[Element] Reals}]
```

returns "y". I will make use of "Simplify" from time to time in the text, but you should always keep this function in mind if you are looking for a particular form of some expression.

One noteworthy case involves the magnitude of a complex expression, which you obtain with the function "Abs" that returns the *absolute value* of the argument. Executing

```
Abs[3 + I 4]
```

returns "5" as you would expect, but

```
z = x + I y
Abs[z]
```

just returns "Abs[x + I y]", whereas you might have been expecting $\sqrt{x^2 + y^2}$. Of course, x or y might themselves be complex, but if you try

```
Simplify[Abs[z], {x > 0 && y > 0}]
```

you get the same result. The problem has to do with deciding which result is more simple. There are some sophisticated ways that you can force MATHEMATICA to decide on simplification criteria, but in this case, it is easiest to use "Complex Expand", which assumes that all variables are real. Indeed,

```
ComplexExpand[Abs[z]]
```

returns the MATHEMATICA equivalent of $\sqrt{x^2+y^2}$.

For more information, see the Wolfram Language Guide and Tutorial on *Complex Numbers* as well as the Tutorials on *Simplification* and *Simplifying with Assumptions*.

2.5 PHYSICS EXAMPLES

Example 2.1 *Train #1 and train #2 move side-by-side on parallel horizontal straight tracks. At time $t = 0$, train #1 starts from rest and accelerates at a constant rate $a_1 = 0.5$ m/s^2. Also at $t = 0$, train #2 passes train #1 while moving at 20 m/s, in the same direction as train #1 is accelerating, but decelerating at a rate $a_2 = 0.2$ m/s^2. Find the time at which train #1 passes train #2, and determine the distance from the start at which they pass.*

Set things up so that position $x = 0$ locates the trains at $t = 0$. That is,

$$x_1 = \frac{1}{2}a_1t^2 \tag{2.1a}$$

$$x_2 = v_0t - \frac{1}{2}a_2t^2 \tag{2.1b}$$

where $a_1 = 0.5$, $a_2 = 0.2$, and $v_0 = 20$. We need to find t when $x_1 = x_2$.

A MATHEMATICA notebook to solve this problem is very simple. My solution is shown in Notebook 2.1. The numerical answer is at the end, namely 816 seconds, or 13.6 minutes. Let's go through this notebook, and pick up a few new concepts.

The first section does just what we learned in this chapter. Note that there are two solutions, since the trains are indeed at the same position at $t = 0$. We use "Part" to extract the time at which the trains cross later. This solution (that is, the replacement statement for the solution) is stored in the variable "tFinite".

The second section checks our answers, by making sure we get the same position for each of the two trains at the later time. At first glance, in fact, the two answers do not appear to be the same. This is because MATHEMATICA does as little as it needs to do when it performs a substitution. In this case, it does not combine the two terms in the equation for $x_2(t)$.

However, we use the built-in MATHEMATICA command "Simplify" to combine the two terms. This command tries a limited number of transformations

NOTEBOOK 2.1 Solution to Example 2.1

```
In[1]:= Remove["Global`*"]
```

Parallel Trains in One Dimension

Define the equations and solve them

```
In[2]:= x1 = (1 / 2) a1 t^2; x2 = v0 t - (1 / 2) a2 t^2;
        soln = Solve[x1 == x2, t]
        tFinite = Part[Part[soln, 2], 1]
```

Out[3]= $\{\{t \to 0\}, \{t \to \frac{2\,v0}{a1 + a2}\}\}$

Out[4]= $t \to \frac{2\,v0}{a1 + a2}$

Check the solutions

```
In[5]:= x1Meet = x1 /. tFinite
```

Out[5]= $\frac{2\,a1\,v0^2}{(a1 + a2)^2}$

```
In[6]:= x2Meet = x2 /. tFinite
```

Out[6]= $-\frac{2\,a2\,v0^2}{(a1 + a2)^2} + \frac{2\,v0^2}{a1 + a2}$

```
In[7]:= Simplify[x2Meet]
```

Out[7]= $\frac{2\,a1\,v0^2}{(a1 + a2)^2}$

Put in the numbers

```
In[8]:= vals = {a1 → 0.5, a2 → 0.2, v0 → 20};
        x1Meet /. vals
```

Out[9]= 816.327

```
In[10]:= % / 60
```

Out[10]= 13.6054

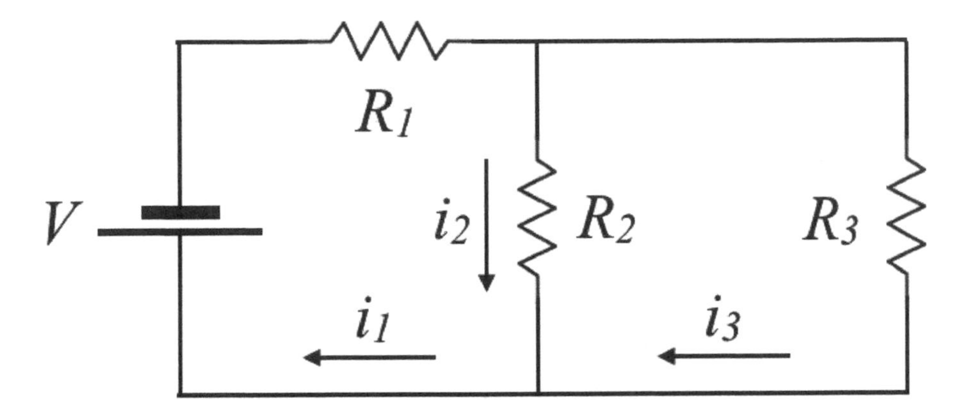

FIGURE 2.1 Figure for Physics Example 2.2.

to make the expression in the argument "as simple as practical." You should consult the MATHEMATICA documentation for more information on simplifying expressions, including the command "FullSimplify".

I find it handy to turn the algebraic expressions into numerical values by first defining a list with the substitutions. This way you can define other lists of substitutions for different numerical cases, which can be useful depending on your application. In this notebook, I use the value from the formula for $x_1(t)$ for determining the position the trains meet, and get 816.327 seconds. (We will discuss significant figures in a later chapter.)

Notice that the notebook uses "%" to divide the answer in seconds by sixty, to get the answer in minutes. The expression "%" contains the last result generated from MATHEMATICA. This is a shorthand for the built-in MATHEMATICA command "Out[1]". You should consult the documentation for variations on this command, as well as other versions of the shorthand.

You'll notice that I used parameters named, for example, $x1$ and $a2$, instead of x_1 and a_2. This is only because, for simple examples, it didn't seem worth the effort to type in subscripts. If you'd like to learn how to do this, check the documentation for a tutorial "Typing Subscripts".

OK, that was pretty easy. Here is a slightly more challenging example.

Example 2.2 *Figure 2.1 shows a DC circuit. Use Kirchoff's Laws to determine the currents i_1, i_2, and i_3 in terms of V, R_1, R_2, and R_3.*

Kirchoff's node and loop laws give the following equations:

$$i_1 = i_2 + i_3 \tag{2.2a}$$
$$V = i_1 R_1 + i_2 R_2 \tag{2.2b}$$
$$0 = i_2 R_2 - i_3 R_3 \tag{2.2c}$$
$$V = i_1 R_1 + i_3 R_3 \tag{2.2d}$$

NOTEBOOK 2.2 Solution to Example 2.2

```
In[1]:= Remove["Global`*"]
```

Current in a DC Circuit

Define the equations

```
In[2]:= eq1 = i1 == i2 + i3;
        eq2 = V == i1 R1 + i2 R2;
        eq3 = 0 == i2 R2 - i3 R3;
        eq4 = V == i1 R1 + i3 R3;
```

Solve the equations

```
In[6]:= Solve[{eq1, eq2, eq3, eq4}, {i1, i2, i3}]
```

$$\text{Out[6]= } \left\{\left\{i1 \to -\frac{-R2\,V - R3\,V}{R1\,R2 + R1\,R3 + R2\,R3},\ i2 \to \frac{R3\,V}{R1\,R2 + R1\,R3 + R2\,R3},\ i3 \to \frac{R2\,V}{R1\,R2 + R1\,R3 + R2\,R3}\right\}\right\}$$

```
In[7]:= Solve[{eq1, eq2, eq3}, {i1, i2, i3}]
```

$$\text{Out[7]= } \left\{\left\{i1 \to -\frac{-R2\,V - R3\,V}{R1\,R2 + R1\,R3 + R2\,R3},\ i2 \to \frac{R3\,V}{R1\,R2 + R1\,R3 + R2\,R3},\ i3 \to \frac{R2\,V}{R1\,R2 + R1\,R3 + R2\,R3}\right\}\right\}$$

Clearly, (2.2d) follows from (2.2b) and (2.2c), but let's come back to that after we review a MATHEMATICA notebook which solves this example.

See Notebook 2.2. The notebook is in fact quit simple. I define four logical statements, calling them "eq1", "eq2", "eq3", and "eq4", and then ask MATHEMATICA to solve them simultaneously for the three currents. I leave the answer as the replacement lists, but you can use these however you'd like, for example to put in numbers, extend the solution to special cases, or apply these solutions to another problem.

You'll note that I checked to be sure I got the same answer if I only use the first three equations. Clearly, MATHEMATICA has no problem with redundant equations in this kind of problem. You might try and see what happens, though, if you make one of the equations inconsistent with the others.

I violated my "good habits" rule of avoiding capital letters to start the names of defined variables, but it's hard not using V for voltage and R for resistances. I was careful when I typed it in, though, to make sure it was blue when I finished the name of the variable. As mentioned in the Introduction, this means that the variable is (so far) unknown to MATHEMATICA.

Finally, one more example, with a few new lessons packed inside.

Example 2.3 *Three point charges are arranged in a straight line. Charges $q_1 = 1$ C, $q_2 = -2$ C, and $q_3 = 4$ C are located at $x_1 = -2$ m, $x_2 = 0$, and $x_3 = 1$ m, respectively. Find the point or points along the line where the electric field is zero.*

Notebook 2.3 shows a solution. We write the electric field from a charge q_i at position x_i as

$$E_i = \frac{q_i}{(x - x_i)^2} \tag{2.3}$$

and add up the three fields. We don't bother with an overall factor $1/4\pi\epsilon_0$ or anything like that, because it is irrelevant for finding out where are the zeros.

This problem is complicated enough so that if we tried to solve it symbolically, without putting in numbers for the positions and charges, it would take a minute or so on a modern computer to come up with the answer, and there would be other complications. (You might want to give it a try and see what happens.)

Here is a good point to make a simple graph of $E(x)$ versus x so that you can see how many zeros should be found, and what are their approximate values. It appears there is a solution close to $x = -1$ and another at $x \approx 0.4$. So, we ask MATHEMATICA to solve the problem. We include the optional argument "Reals" so that only the domain of real numbers is searched. Since the resulting algebraic equation will be some higher degree polynomial, it is likely that we would find complex solutions, but those, of course, do not represent this physical situation.

You'll notice that I didn't bother to let the symbolic result print. (You may try doing so yourself, just remove the semicolon at the end of the "Solve" command, and see what happens.) Instead, I use "N" to immediately reduce the solution to numbers. Nevertheless, the result is a little surprising.

"Solve" indeed returns two solutions (in a list) which we return in a list of "numbers" and store in the variable "xVal". The first solution is, indeed, $x = -1$. The second argument,

```
Root[-8 + 16 #1 + 7 #1^2 + 3 #1^3 &, 1]
```

looks peculiar, however. You might look up "Root" in the MATHEMATICA documentation, but, essentially, it represents the exact value of the root of the polynomial equation that is its first argument, and the second argument tells which root you want, in this case the first (and only) root.

The syntax is peculiar, with the ampersand "&" signaling this as a "pure function" and "#1" representing the one argument. In other words, "Root" returns the one real solution x of the equation $-8+16x+7x^2+3x^3 = 0$. This is the quantity that is turned into the real number 0.412429 by the function "N" .

2.6 CHAPTER SUMMARY

- Equations are logical expressions. Familiarize yourself with the double equals "==" sign.
- Familiarize yourself also with the concept of a *list* and basic list manipulations, including "Part".

NOTEBOOK 2.3 Solution to Example 2.3

```
In[1]:= Remove["Global`*"]
```

Electric Field from Three Charges

Define positions and charges

```
In[2]:= vals = {x1 → -2, x2 → 0, x3 → 1,
           q1 → 1, q2 → -2, q3 → 4};
```

Find the fields from the three charges

```
In[3]:= Efield1 = q1 / (x - x1) ^2;
        Efield2 = q2 / (x - x2) ^2;
        Efield3 = q3 / (x - x3) ^2;
        Efield = Efield1 + Efield2 + Efield3;
        EfieldVals = Efield /. vals;
```

```
In[8]:= Plot[EfieldVals, {x, -3, 2}]
```

Out[8]=

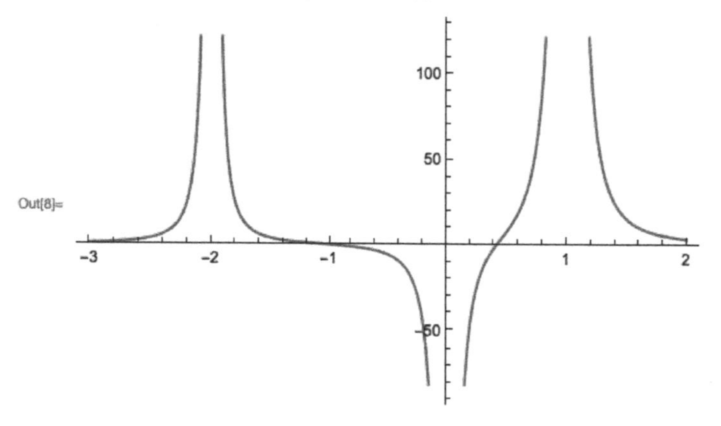

Find the positions where the field is zero

```
In[9]:= soln = Solve[EfieldVals == 0, x, Reals];
        xVals = x /. soln
        N[xVals]
```

Out[10]= $\{-1,\ \text{Root}[-8 + 16\,\#1 + 7\,\#1^2 + 3\,\#1^3\ \&,\ 1]\}$

```
Out[11]= {-1., 0.412429}
```

- The function "Solve" returns a double-nested list of replacement statements. The outer list is because there can be more than one solution. The inner list(s) allow for solutions for more than one variable.
- MATHEMATICA naturally allows for imaginary and complex numbers.
- The percent sign "%" is a shortcut for the function "Out" which gets the last output result.
- The function "Root" finds the root of a function.

EXERCISES

2.1 *A projectile is launched with initial speed v_0 and at an angle θ with respect to the horizontal. Its position (x, y) at any time t is given by*

$$x = v_0 t \cos\theta \qquad \text{and} \qquad y = v_0 t \sin\theta - gt^2/2$$

where x (y) measures the horizontal (vertical) position and g is the acceleration due to gravity near the Earth's surface. Solve for the time(s) at which $y = 0$, and use this to determine the range, that is, the horizontal distance at which the projectile hits the ground.

2.2 *See Exercise 2.1, but now the projectile is fired from the edge of a cliff. The cliff is a height $y = h$ above the ground, so the motion is described by*

$$x = v_0 t \cos\theta \qquad \text{and} \qquad y = h + v_0 t \sin\theta - \frac{1}{2} g t^2$$

Find the range of the projectile. You might check that you get the answer from Exercise 2.1 for $h = 0$.

2.3 *A simple harmonic oscillator moves in one dimension x with velocity v where*

$$\begin{aligned} x(t) &= A\cos(\omega t + \phi) \\ \text{and} \qquad v(t) = \dot{x}(t) &= -\omega A \sin(\omega t + \phi) \end{aligned}$$

If at time $t = 0$ the initial position is $x(0) = x_0$ and the initial velocity is $v(0) = v_0$, find A and ϕ in terms of x_0 and v_0.

2.4 *An electron is accelerated through a potential difference V and then moves with constant velocity along the z-axis, when it encounters a region of crossed, uniform electric and magnetic fields. The electric field is in the y-direction with magnitude 1 kV/cm (100 kV/m), and the magnetic field is in the x-direction with magnitude 0.025 T. What must be the value of V so that the electron continues through the crossed-field region at the same velocity?*

2.5 *In plane polar coordinates (r, ϕ), orbits around the Sun follow the path*

$$r(\phi) = \frac{r_0}{1 + \epsilon \cos\phi}$$

One asteroid follows a path with $r_0 = 15$ and $\epsilon = 1/8$ and a second asteroid follows a path with $r_0 = 14$ and $\epsilon = 1/4$. Find the position(s) at which the two paths intersect.

Suggestion: *Use this exercise to explore some slightly advanced features of* MATHEMATICA. *Solve the exercise generally by finding the angle(s) ϕ at which two orbits $r_1(\phi)$ and $r_2(\phi)$ intersect. Express $r_1(\phi)$ and $r_2(\phi)$ in terms of two scale parameters a_1 and a_2 and eccentricities ϵ_1 and ϵ_2. You will find a "conditional expression" that stems from the fact that ϕ can be multivalued by adding multiples of 2π. Experiment with "*Normal*" and replacement ("/.") and "*Part*" to extract a single, nominally positive expression for the intersection angle. Then, finally, insert the numerical values; the numbers I gave you seem peculiar, but the answer turns out to be a simple angle.*

2.6 *A basketball is launched from mid-court at a 30° angle with the horizontal and at a height of 2 m towards the basket 12.5 m away. If the rim is 3 m above the floor, neglecting air resistance, what must be the initial speed of the ball in order to sink the basket?*

2.7 *The figure below shows a resistor R, and inductor L, and a capacitor C connected in parallel:*

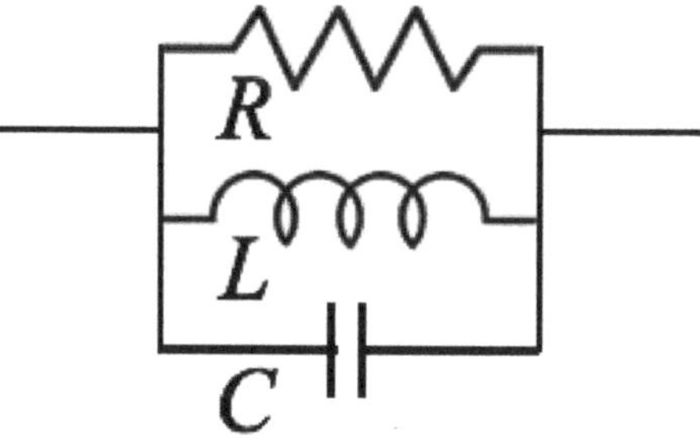

An AC potential difference $V(t) = V_0 \cos\omega t = \mathrm{Re}(e^{i\omega t})$ is applied across the open terminals. In this case, the concept of resistance *for DC circuits can be replaced by* impedance *for AC circuits, where resistive impedance is $Z_R = R$, inductive impedance is $Z_L = i\omega L$, and capacitive impedance is $Z_C = 1/i\omega C$. The Z's can be used as resistances using the same rules for DC circuits, but the complex result has both a magnitude and a phase, as defined by the absolute value ("*Abs*" in* MATHEMATICA*) and phase ("*Arg*").*

Find an expression for the impedance Z of the above circuit segment, in terms of R, L, C, and the frequency ω. Using the realistic circuit values $R = 1$ kΩ, $C = 100$ μμF, and $L = 5$ mH, plot the magnitude and phase *of Z as a function of ω. It is best to make the plot using a log scale for ω; see the function* "LogLinearPlot" *which will do this for you.*

Now solve for the value of ω at which the phase passes through zero. You might encounter difficulties using "Solve", *depending on how you determined the phase, because of the way* MATHEMATICA *treats complex numbers. However, using* "Refine" *with the assumptions that R, L, C, and ω are all real and positive, combined with* "Re", "Im", *and* "ComplexExpand", *you can end up with a brute force version of the phase that will work with* "Solve".

CHAPTER 3

Derivatives, Integrals, and Series

CONTENTS

As physicists, we study things that vary continuously. This could be the position of a projectile as it varies with time, the quantum mechanical wave function of a particle as it varies through space, or the temperature of a heated conducting plate as it varies in both position and time.

For this reason, calculus is essential to the "language" of physics. Calculus tells us how to treat the instantaneous change of something which varies continuously. It also tells us how to sum up quantities that are never constant.

This is why MATHEMATICA is so useful for studying physics. Physically, a derivative is the ratio of two very tiny, that is infinitesimal, quantities. An integral is a sum over a very large, in fact infinite, number of infinitesimal quantities. From the point of view of physics, we formulate the solutions to problems this way, and MATHEMATICA can take over the task of actually performing the mathematics needed to evaluate the derivatives and integrals.

This chapter takes you through the steps of calculating derivatives and integrals. We will also learn how to perform Taylor series expansions, and apply the results to approximations where certain variables may be very small or very large. Many important functions cannot be integrated analytically, so we also discuss how to carry out numerical integration.

3.1 DERIVATIVES

You can take the derivative of an expression, or the derivative of a function. Let's start with expressions.

The function "D[f,x]" takes the (first) derivative with respect to x of the expression $f(x)$. You can take the second derivative with "D[f,{x,2}]" and the nth derivative with "D[f,{x,n}]". So, for example

```
D[x^2, x]
D[x^a, {x, 2}]
```

returns the lines "2x" and "(-1+a)ax^{-2+a}".

In fact, "D[f,x]" returns the partial derivative $\partial f/\partial x$, which of course is simply the derivative for expressions that are functions $f(x)$ of a single variable. However, "D[f,x,y]" returns $\partial^2 f/\partial x\partial y$ and "D[f,{x,2},{y,3}]" returns $\partial^5 f/\partial x^2\partial y^3$ for an expression $f(x,y)$. For example,

```
f = Exp[-a t] Sin[Pi x];
D[f, {x, 2}, t]
```

returns "ae$^{-at}\pi^2$Sin[πx]".

Note that in the previous example, we implicitly took the derivative of the functions "Exp" and "Sin". Indeed, "D[Cos[x], x]" returns "-Sin[x]". You get the same result if you use "Cos'[x]" instead of "D[Cos[x], x]"; that is, you can use the "prime" notation for derivative of a function of single variable. You should try "Cos''[x]" and see what you get.

This all works for user defined functions, too, of course. In fact, if you enter "D[func[z], z]" then MATHEMATICA returns "func'[x]" if "func" is undefined. However, the statements

```
func[var_] = var  Exp[2 var];
D[func[z], z]
```

return "e^{2z} + 2e^{2z} z". I'll remind you again, however, that I do not recommend using functions in place of expressions until you've gotten the hang of things and realize that a function is in fact called for in your particular problem.

MATHEMATICA is aware of many special functions and their derivatives. You are likely to come across the "Heaviside" function $\theta(x)$ which is zero for $x < 0$ and unity for $x > 1$. Its derivative is the "Dirac delta-function" $\delta(x)$, which you have probably seen as well. Indeed, "D[HeavisideTheta[x], x]" returns "DiracDelta[x]".

Physics applications will often make use of derivative vector calculus with operators like gradient, divergence, curl, and the Laplacian. These are all built in to MATHEMATICA as well. I recommend reading up on these in the documentation. There is a language guide and also a tutorial on "Vector Analysis" in MATHEMATICA. I've also included a brief introduction to these functions in Section 12.4.

3.2 INDEFINITE INTEGRALS

The indefinite integral, or antiderivative, of an expression or function is just as simple to take as a derivative. The command is the verb "Integrate". (Apparently, we should read "D" as "differentiate".) The syntax is the same, that is "Integrate[2 x, x]" returns "x^2", "Integrate[4 x y, x, y]" returns "x^{2}y^2" (but you cannot integrate on a variable more than once), and "Integrate[Cos[x], x]" returns "Sin[x]".

Of course, MATHEMATICA can handle special functions. For example,

```
Integrate[1/x, x]
```

returns "Log[x]",

```
Integrate[DiracDelta[x], x]
```

returns "HeavisideTheta[x]",

```
Integrate[HeavisideTheta[x], x]
```

returns "x HeavisideTheta[x]", and

```
(2/Sqrt[Pi]) Integrate[Exp[-x^2], x]
```

returns "Erf[x]", aka, the "error function."

Technically, functions like "Exp[-x^2]" have no antiderivative, but because the integral is so useful, it is given a special name. (For another example, see what you get with "Integrate[Sin[x]/x, x]".) MATHEMATICA will try very hard to express your integral in terms of known special functions, but sometimes there is just no place to go. For example, try

```
Integrate[1/Sqrt[1 - Cos[Cos[x]^2]], x]
```

and MATHEMATICA will just give you

$$\int \frac{1}{\sqrt{1-\cos\left(\cos^2(x)\right)}}\, dx$$

That is, the output is the same expression as the input.

Let's pause for a moment to introduce the notion of output forms in MATHEMATICA. This book is written in LaTeX, and I was able to produce the above expression using "TeXForm[%]", immediately following the "Integrate" command. That is, MATHEMATICA gave me the output format in such a way that I could just paste it into the script I am using for this chapter.

There are a number of other output forms that we'll see along the way, but for now, I just wanted you to know that they were in there.

3.3 DEFINITE INTEGRALS

Use "Integrate" to compute definite integrals, but now the second argument is a list that includes the lower and upper limits as well as the integration variable. For example

```
Integrate[1/x, {x, 1, E^2}]
```

returns the value "2", and

```
Integrate[2 Cos[x] Sin[x], {x, 0, z}]
```

returns the expression "Sin[z]^2".

The MATHEMATICA built-in object called "Infinity" has an obvious meaning and usage. So, for example,

```
Integrate[2 Exp[-x^2], {x, 0, Infinity}]
```

returns the value "$\sqrt{\pi}$".

It often happens that parameters in the integrand expression will impose constraints on the result. For example,

```
Integrate[A x^n, {x, 0, 2}]
```

returns a "conditional expression" because the integrated expression, that is $Ax^{1+n}/(1+n)$, can only be evaluated at $x = 0$ if $n > 1$. However, MATHEMATICA allows you to apply "assumptions" to parameters when carrying out a variety of calculations, including "Integrate". In this case

```
Integrate[A x^n, {x, 0, 2}, Assumptions -> n > 0]
```

simply returns "2^{1+n}A/(1+n)".

This might be a good time for you to look up the MATHEMATICA language guide on *Assumptions and Domains.*

3.4 NUMERICAL INTEGRATION

The MATHEMATICA command "NIntegrate" allows you to carry through an integration numerically, using the same format as for definite integrals with "Integrate". That is, the integral is carried out using some numerical algorithm that divides up the integration region into small pieces and adds them up. Therefore, if you try something very simple like

```
NIntegrate[x, {x, 0, 1}]
```

you get "0.5" for the answer. This is not the same as $\frac{1}{2}$, but instead a *floating point* number, to use some computer science jargon.

Many important integrals in physics cannot be done analytically, so numerical integration can be very useful. For example, the integral of a Gaussian is frequently encountered in statistical analysis, and you might need something like

```
NIntegrate[Exp[-x^2], {x, 0, 0.5}]
```

which returns "0.461281". Of course, you can get the same answer using "(Sqrt[Pi]/2) Erf[0.5]" since

$$\mathrm{Erf}(z) = \frac{2}{\sqrt{\pi}} \int_0^z e^{-t^2} dt$$

.

Indeed, if an integral can't be done analytically but it still has important physical significance, history has likely assigned it the name of some special function known to MATHEMATICA.

3.5 POWER SERIES

Physics solutions frequently make use of approximations that rely on some quantity being "small" with respect to some other quantity. These approximations seek to find expansions in powers of this small quantity, and then truncate the expansion to the desired level of precision.

Taylor's theorem, from mathematics, tells us that any differentiable function can be expressed in a power series expansion about some point. In MATHEMATICA, this is realized by the command "Series" which allows an arbitrary function to be written this way, up to a specified order. For example,

```
Series[(1 + x)^a, {x, 0, 2}]
```

returns the expansion of $(1+x)^a$ about $x=0$ up to second order, that is

$$1 + \mathrm{ax} + \frac{1}{2}\left(-\mathrm{a} + \mathrm{a}^2\right)\mathrm{x}^2 + \mathrm{O}\left[\mathrm{x}^3\right]$$

Note that this is a special output, including the "$\mathrm{O}\left[\mathrm{x}^3\right]$" at the end. If you do operations with power series in this form, MATHEMATICA keeps the order consistent. On the other hand, if you just want the approximate expression up to that order, use "Normal". For example,

```
pow = Series[(1 + x)^a, {x, 0, 2}];
Normal[pow]
```

returns just

$$1 + \mathrm{ax} + \frac{1}{2}\left(-\mathrm{a} + \mathrm{a}^2\right)\mathrm{x}^2$$

"Normal" is useful for converting various different special forms, which we will discuss as needed in this book.

3.6 PHYSICS EXAMPLES

Example 3.1 *A particle of mass m moves in one dimension x according to*

$$x(t) = Ae^{-\beta t}\cos\omega t$$

NOTEBOOK 3.1 Solution to Example 3.1

```
In[1]:= Remove["Global`*"]
```

The damped oscillator

Construct position, velocity, acceleration

```
In[2]:= x = A Exp[-β t] Cos[α t] /. α → Sqrt[ω^2 - β^2];
        v = D[x, t];
        a = D[v, t];
```

Now construct the force and compare to "ma"

```
In[5]:= f = -k x - b v /. {k → m ω^2, b → 2 m β}
```

$$\text{Out[5]= } -A\, e^{-t\beta}\, m\, \omega^2 \cos\left[t\sqrt{-\beta^2+\omega^2}\right] - 2\, m\, \beta \left(-A\, e^{-t\beta}\, \beta \cos\left[t\sqrt{-\beta^2+\omega^2}\right] - A\, e^{-t\beta}\sqrt{-\beta^2+\omega^2}\, \sin\left[t\sqrt{-\beta^2+\omega^2}\right]\right)$$

```
In[6]:= Simplify[f - m a]
Out[6]= 0
```

Show that this motion corresponds to the mass being acted on by a restoring force $-kx$ and a damping force $-bv$, where v is the velocity, $\omega = \sqrt{k/m}$ and $\beta = b/2m$.

We will be getting on to actually solving differential equations soon, but for now we just want to show that the given expression *is* a solution to the equation $ma = m\ddot{x} = f = -kx - b\dot{x}$.

The solution shown in Notebook 3.1 is a straightforward one. After construction of an expression for $x(t)$ we take its derivative once for the velocity, and twice for the acceleration. We then construct the force, and as shown, we print the output form here. The expression appears rather complicated and messy. However, when we subtract "mass×acceleration", and ask MATHEMATICA to simplify the result, we indeed get zero. (There are cleaner ways to check the solution, using logical expressions. We'll discuss these in Section 6.2.)

Example 3.2 *Consider a pendulum made from a massless string of length l and a bob of mass m. We generally write the period as $2\pi\sqrt{l/g}$ but this is accurate only if the maximum displacement angle $\theta_0 \ll 1$. Find an expression (perhaps in terms of a special function) for the period as a function of θ_0. Form the ratio of the period to $2\pi\sqrt{l/g}$ and plot it for $0 \leq \theta_0 \leq 0.99\pi$.*

We can find an integral for the period by using conservation of energy.

Equate the general expression for the mechanical energy, to the potential energy when the pendulum is at its highest point ($\theta = \theta_0$). That is

$$\frac{1}{2}m\left[\frac{d}{dt}(l\theta)\right]^2 + mgl(1-\cos\theta) = mgl(1-\cos\theta_0)$$

Reduce this to find an expression for the angular velocity $d\theta/dt$:

$$\frac{d\theta}{dt} = -\sqrt{\frac{2g}{l}}\sqrt{\cos\theta - \cos\theta_0}$$

We chose the minus sign because we'll be thinking about the quarter period of the pendulum after it is released from rest at $\theta = \theta_0$, when the angle is decreasing.

The period T is four times the time it takes for the pendulum to rotate from its initial resting point at $\theta = \theta_0$, to vertical, that is $\theta = 0$. Mathematically, this is written as

$$T = 4\int_{\theta_0}^{0} dt = 4\int_{\theta_0}^{0} \frac{d\theta}{d\theta/dt} = -4\sqrt{\frac{2g}{l}}\int_{\theta_0}^{0} \frac{d\theta}{\sqrt{\cos\theta - \cos\theta_0}}$$

This looks like a scary integral, and indeed there is no closed analytic form for it, but it is no problem for MATHEMATICA.

Notebook 3.2 shows a solution. It takes my computer a little while to work on the integral, so included here is a way to record how much time is used. The "AbsoluteTime[]" function returns the time in seconds from a specific date far in the past. By using this just before and just after the function call to "Integrate", we determine the time spent inside that routine. Subtracting the two values, and dividing by 60, tells us that it spends 0.204 minutes to execute the "Integrate" command for this function.

Notice that a logical statement is included as "Assumptions" for the integral. This statement simply says that $0 < \theta_0 < \pi$. These limits ensure that the pendulum is in just one quadrant. The upper limit is important because the integral will diverge for $\theta_0 = \pi$. (The pendulum will just rest in its vertical equilibrium position forever and never oscillate. That is, the period is infinite.)

The answer is indeed in terms of a special function, "EllipticF". This is a so-called elliptic integral of the first kind. The details are not important for this discussion, you only need to know that MATHEMATICA knows how to evaluate it internally. If you'd like, you can look through the MATHEMATICA documentation to learn more about elliptical integrals and similar constructs.

I decided to plot the ratio (of the period to $2\pi\sqrt{l/g}$) using "LogPlot" instead of "Plot". The syntax is the same, but the vertical axis is logarithmic. This makes it possible to see the small variation for small angles, and also the extreme values when the angle is larger than 2 radians or so.

You might consider how to make this plot against $\theta_0/2\pi$ instead of θ_0.

Here is an example that makes use of power series for an approximation.

NOTEBOOK 3.2 Solution to Example 3.2

```
In[1]:= Remove["Global`*"]
```

Pendulum period

Do the integral and get the period

```
In[2]:= t1 = AbsoluteTime[];
        int = Integrate[1 / Sqrt[Cos[θ] - Cos[θ0]], {θ, θ0, 0},
            Assumptions → {θ0 > 0 && θ0 < Pi}];
        t2 = AbsoluteTime[];
        time = (t2 - t1) / 60
Out[5]= 0.20494423

In[6]:= period = -4 Sqrt[1 / (2 g)] int
```

Out[6]=

$$\frac{4\sqrt{2}\sqrt{\frac{1}{g}}\,\text{EllipticF}\left[\frac{\theta 0}{2},\,\text{Csc}\left[\frac{\theta 0}{2}\right]^2\right]}{\sqrt{1-\text{Cos}[\theta 0]}}$$

Form the ratio and plot it

```
In[7]:= ratio = period / (2 Pi Sqrt[1 / g]);
In[8]:= LogPlot[ratio, {θ0, 0, 0.99 Pi}]
```

Out[8]=

NOTEBOOK 3.3 Solution to Example 3.3

```
In[1]:= Remove["Global`*"]
```

Gravity near the Earth's surface

Define the force of gravity

```
In[2]:= force = G m M / (R + h) ^2;
```

Substitute and expand

```
In[3]:= Series[force /. h → x R, {x, 0, 2}];
        forcex = Normal[%] /. x → h / R
```

Out[4]= $\frac{3 G h^2 m M}{R^4} - \frac{2 G h m M}{R^3} + \frac{G m M}{R^2}$

```
In[5]:= grepl = Solve[g == G M/R^2, G];
        forcex /. grepl
```

Out[6]= $\left\{g m + \frac{3 g h^2 m}{R^2} - \frac{2 g h m}{R}\right\}$

```
In[7]:= Simplify[%]
```

Out[7]= $\left\{\frac{g m \left(3 h^2 - 2 h R + R^2\right)}{R^2}\right\}$

Example 3.3 *At the Earth's surface, we speak of the "acceleration g due to gravity". In terms of Newton's theory and a spherical Earth of radius R and mass M,* $g = GM/R^2$. *Find an expression up to second order for the modification to g for an object at height h.*

One of the things to realize is that only dimensionless quantities can be "small", so we need to identify the small quantity in this case. In this problem, it is rather obvious that the small quantity is h/R. Therefore, we write

$$F = G\frac{mM}{(R+h)^2} \qquad \text{with} \qquad h = xR$$

so that x is small.

Notebook 3.3 gives one solution to carrying out this approximation. After defining the force, we do a series expansion on x about $x = 0$, to second order, after making the replacement $h = xR$. Then, after converting to normal form, we undo the replacement to get the expansion in terms of h instead of x.

However, we are asked to express the answer as a modification of g, so there is more work to do. Although it is simple to do on paper, this solution solves for G in terms of g, and then replaces that form in the force. After simplification, it is quite clear that our final answer would be written as

$$g(h) = g_0 \left(1 - 2\frac{h}{R} + 3\frac{h^2}{R^2}\right)$$

where $g_0 \equiv GM/R^2$.

One thing that should be clear from this example, is that although you can take the output as far along as you'd like on MATHEMATICA, it is of course reasonable to get it close enough so that you can write down the solution and be done with it. On the other hand, especially if you are making presentations, you may want to have the output looking prettier. We will leave this kind of example and exercise for later.

3.7 CHAPTER SUMMARY

- Take (partial) derivatives of expressions or functions with the function "D".
- Indefinite integrals are obtained with "Integrate".
- MATHEMATICA understands entire classes of special functions.
- Use "Integrate" with integration limits to obtain definite integrals. Sometimes, conditional expressions may result.
- Numerical integration is carried out using "NIntegrate".
- Power series expansions are straightforward using "Series". You may want to follow up with "Normal" to get a normal expression that does not include the *order* indicator.
- The function "AbsoluteTime" can be used to determine elapsed time while running a notebook.

EXERCISES

3.1 *An object moves in one dimension according to*

$$x(t) = 5te^{-t^2}$$

Find the velocity and acceleration, and plot these along with the position on a single graph with legends. Find the time(s) at which there is no net force on the object.

3.2 *An object moves in one dimension according to*

$$x(t) = (1+t)e^{-2t} - (1-t)$$

Show that this motion results from a force $F(t) = 2te^{-2t}$ applied to a particle of mass $m = 1/2$ whose initial position and velocity are both zero.

3.3 *Orbits in plane polar coordinates* (r, ϕ) *are described by*

$$r(\phi) = \frac{r_0}{1 + \epsilon \cos \phi}$$

which is an ellipse for $\epsilon < 1$. *The semi-major axis* a *of the ellipse is one half the distance between the minimum and maximum values of* r. *Set* $dr/d\theta = 0$ *to determine* a *in terms of* r_0 *and* ϵ.

3.4 *Calculate the electric field along the axis of a thin disk of radius* R *and charge* Q *lying in a plane, at a distance* z *from the plane of the disk. Show that you get the answer you expect in the limits* $z \ll R$ *and* $z \gg R$. *(Hint: You may find it useful to put assumptions on* z *and* R *when evaluating the definite integral.)*

3.5 *A lifeguard runs from a fixed point on the beach "A" to a spot on the shore, then swims in the ocean "B" to reach a bather:*

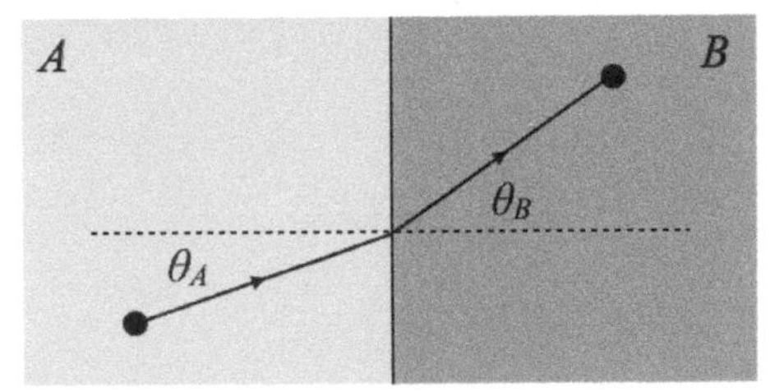

Her speed is v_A *on the sand and* v_B *in the water. Find a relationship between the angles* θ_A *and* θ_B *that minimizes the time it takes for the lifeguard to reach the bather. (Hint: Define a coordinate* x *that runs along the shore, and minimize the time with respect to* x. *You'll need some creative use of "*ReplaceAll*" to cast the resulting equation in terms of the angles.)*

3.6 *Referring back to Exercise 1.6, calculate the potential energy* $U(x)$ *of the charge* Q *by integrating the force, for the case where* Q *and* q *have opposite signs. Expand* $U(x)$ *about* $x = 0$ *to show that* $U(x)$ *is proportional to* x^2 *if* $x \ll a$.

CHAPTER 4

Differential Equations: Analytic Solutions

CONTENTS

The laws of physics are formulated as differential equations. Newton's Second Law $F = ma$ is a second order differential equation for $x(t)$ with $a = d^2x/dt^2$ and $F = F(x, dx/dt, t)$. Maxwell's Equations, written in differential form, are coupled partial differential equations for the electric and magnetic fields. Heat conduction, diffusion, and Schrödinger's Equation are further examples. In formulating the solutions to most problems, we end up with a differential equation that needs to be solved.

This is a place where MATHEMATICA shines for physics students. The "physics" is in the formulation of the problem as a differential equation, and this follows from general physical principles. Solving the differential equation, however, can be difficult or even impossible unless one resorts to numerical approaches. The solution is nevertheless essential, and MATHEMATICA provides that solution straightforwardly.

The primary MATHEMATICA command you'll use for finding analytic solutions of differential equations is "DSolve". We'll discuss numerical solutions in the next chapter.

If you immediately turned to this chapter, I don't blame you. In fact, I will try to refer back to earlier material more frequently in this chapter, to help you fill in the blanks if you were impatient.

4.1 FIRST ORDER ORDINARY DIFFERENTIAL EQUATIONS

Let's begin with a simple example, namely solving $df/dx = af(x)$. There are two ways in which you can get "DSolve" to return the solution, and we'll follow both of them here. With practice, and depending on the situation, you may prefer one or the other.

The first type of solution comes by entering

```
DSolve[f'[x] == a f[x], f, x]
```

for which MATHEMATICA returns

```
{{f -> Function[{x}, E^(a x) C[1]]}}
```

The second type of solution uses "f[x]" in the second argument that is

```
DSolve[f'[x] == a f[x], f[x], x]
```

for which MATHEMATICA returns

```
{{f[x] -> E^(a x) C[1]}}
```

The right answer, i.e. $f(x) = Ce^{ax}$, is clearly returned in each case, but let's look carefully at what we entered and what came out.

The syntax of "DSolve" is very similar to "Solve", which we discussed in Chapter 2. The first argument is the differential equation (or list of equations) that you want to solve. The second argument identifies the function that is the solution you're looking for, and the third argument is the dependent variable of that function.

Also as with "Solve", "DSolve" returns a nested list with a replacement statement. The first way of using "DSolve" replaces the solution target function "f" with a *pure function* "Function". You should look up "Function" in the MATHEMATICA documentation, but, in its simplest form, it is just a formal way to write a function. For example

```
Function[w,w^2+w]
```

is just the function $f(w) = w^2 + w$ and can be used just that way. In other words

```
Function[w, w^2 + w][y]
```

returns the expression

```
y + y^2
```

The form of "DSolve" that uses "f[x]" as the target function returns a simpler, but less flexible, solution, namely a replacement for the function "f[x]". I will tend to use this simpler form in the rest of this book.

Let's return to the replacement statement that is the solution to our simple differential equation. It clearly replaces "f" with a pure function that is just

e^{ax} times a constant, but let's try to understand the form "C[1]" of that constant. It is capitalized, so it must be a built-in MATHEMATICA function. (Recall the discussion on page 3.) The MATHEMATICA documentation will tell you "C[i] is the default form for the ith parameter or constant generated in representing the results of various symbolic computations." That's clear enough. You can just go ahead and replace "C[1]" with a value or parameter, that is

```
sol = DSolve[f'[x] == a f[x], f[x], x];
sol /. C[1] -> b
```

will return

```
{{f[x] -> b E^(a x)}}
```

In physics, of course, we are generally solving differential equations that have initial or boundary conditions, and "C[1]" in our simple example is just holding the place of such a condition. It is better practice, therefore, to include the boundary condition in the "DSolve" statement by using a list of equations for the first argument. For example,

```
sol = DSolve[{f'[x] == a f[x], f[0] == b}, f[x], x]
```

returns

```
{{f[x] -> b E^(a x)}}
```

(If you jumped to this chapter, now would be a good time to review the use of lists in MATHEMATICA. See Section 2.2 and then the relevant documentation.)

Now let's see how to use this replacement statement. Starting from above, it is clear that "Part[Part[sol, 1], 1]" returns just the statement, out of the nested lists. So, if you simply want to come up with an expression that is the function $y(x)$ that is the solution to the differential equation and boundary conditions, you can do something like

```
sol = DSolve[{f'[x] == a f[x], f[0] == b}, f[x], x];
solAlone = Part[Part[sol, 1], 1];
y = f[x] /. solAlone;
```

and the variable y will be equal to be^{ax}. You can of course choose an independent variable other than x in the line "y = f[x] /. solAlone;" above.

Finally, we note that you don't have to use the "prime" notation for derivatives. The differential equation in the above example can also be written

```
D[f[x], x] == a f[x]
```

and you will get the same result.

If this is the first section of the book you've read, you should consider skipping now to Example 4.1 in Section 4.5. This is a full physics problem, worked out using most of the material discussed up to this point.

4.2 SECOND ORDER ORDINARY DIFFERENTIAL EQUATIONS

There is no particular complication to moving to second order equations. The syntax and solution format is the same as for first order. So, if you enter

```
sol = DSolve[f''[x] == -k^2 f[x], f, x]
```

you get in return

```
{{f -> Function[{x}, C[1] Cos[k x] + C[2] Sin[k x]]}}
```

and the rest of the lessons of Sec. 4.1 apply. Notice that you can specify the second order boundary conditions in the analogous way, that is, executing

```
sol = DSolve[{f''[x] == -k^2 f[x], f[0] == a, f'[0] == b}, f, x]
```

returns

```
{{f -> Function[{x}, (a k Cos[k x] + b Sin[k x])/k]}}
```

Since "$F = ma$" is in fact a second order differential equation for x (where $a = \ddot{x}$), this opens up a world of mechanics problems that can be addressed with MATHEMATICA.

4.3 SIMULTANEOUS DIFFERENTIAL EQUATIONS

Just as "Solve" can be used to find solutions to simultaneous systems of algebraic equations (see Sec. 2.3), we can use "DSolve" to do the same for differential equations. The syntax is just what you would expect, based on what you've already seen.

For example, to solve the equations

$$\frac{d^2y}{dx^2} = z \qquad \text{and} \qquad \frac{dz}{dx} = y$$

subject to the initial conditions $y(0) = 1$, $y'(0) = 0$, and $z(0) = -1$, try

```
sol = DSolve[
   {f''[x] == g[x], g'[x] == f[x],
    f[0] == 1, f'[0] == 0, g[0] == -1},
   {f, g}, x];
```

and then find expressions for y and z with

```
y = f[x] /. Part[Part[sol, 1], 1]
z = g[x] /. Part[Part[sol, 1], 2]
```

Solutions to simultaneous differential equations are generally not obvious. In this case, you find

$$y = \frac{1}{3}e^{-x/2}\left(\sqrt{3}\sin\left(\frac{\sqrt{3}x}{2}\right)+3\cos\left(\frac{\sqrt{3}x}{2}\right)\right)$$

and

$$z = -\frac{1}{3}e^{-x/2}\left(3\cos\left(\frac{\sqrt{3}x}{2}\right)-\sqrt{3}\sin\left(\frac{\sqrt{3}x}{2}\right)\right)$$

I used "TeXForm" to get these equations in LaTeX form from my notebook.

4.4 PARTIAL DIFFERENTIAL EQUATIONS

You can use "DSolve" to solve partial differential equations. All that matters is that the equation you ask it to solve is, in fact, a partial differential equation for a function of two variables.

Unfortunately, however, partial differential equations (without boundary conditions) do not generally have unique solutions. For physical problems, it is best to work out the solutions on a case-by-case basis.

Consider, for example[1], the partial differential equation

$$x\frac{\partial u}{\partial x} - 2\frac{\partial u}{\partial y} = 0 \tag{4.1}$$

to be solved for the function $u(x, y)$. You would solve this in exactly the same way that we've solved ordinary differential equations earlier in this chapter, only to take care to specify that we are looking for a function of two variables, and to take the derivatives of the appropriate ones. That is, you would enter the commands

```
pde = x D[f[x, y], x] - 2 y D[f[x, y], y] == 0;
sol = DSolve[pde, f[x, y], {x, y}];
u = f[x, y] /. Part[Part[sol, 1], 1]
```

in which case MATHEMATICA returns for "u" the expression "C[1][x^{2}y]". Note something similar to, but very different from, our first example with an ordinary differential equation. The symbolic parameter "C[1]" (see Page 45) this time is a *function*. That is, the solution to Equation 4.1 is an arbitrary function of the combination x^2y, i.e. $u(x, y) = g(x^2y)$ where $g(z)$ is an arbitrary (differentiable) function of z.

For ordinary differential equations, we used boundary (or initial) conditions to remove the arbitrary parameters, so we should try something similar here. Suppose we specify that the solution satisfy $u(1, y) = 2y + 1$. Then we would solve this with

[1]This example is taken from *Mathematical Methods for Physics and Engineering, 3rd Ed.*, by K. F. Riley, M. P. Hobson, and S. J. Bence, Cambridge University Press (2006).

```
pde = x D[f[x, y], x] - 2 y D[f[x, y], y] == 0;
sol = DSolve[{pde, f[1, y] == 2 y + 1}, f[x, y], {x, y}];
u = f[x, y] /. Part[Part[sol, 1], 1]
```

and discover that $u(x, y) = 1 + 2x^y$, a unique solution with no arbitrary functions or parameters.

However, if we specify instead that $u(1, 1) = 4$, that is use

```
sol = DSolve[{pde, f[1, 1] == 4}, f[x, y], {x, y}]
```

then we discover that MATHEMATICA cannot return a solution. Indeed, it is pretty clear to see that there is no unique solution or even a unique form of a solution. Each of the following expressions

$$\begin{aligned} u(x, y) &= x^2y + 3 \\ u(x, y) &= 4x^2y \\ u(x, y) &= 4 \end{aligned}$$

solves the differential equation and $u(1, 1) = 4$.

When working with partial differential equations for physics problems, we can rely on nature giving us a unique solution when it makes sense, physically, that one should exist.

4.5 PHYSICS EXAMPLES

Example 4.1 *A mass m is fired vertically upward from the Earth's surface with an initial speed v_0. The mass is subject to a drag force bv, that is, proportional to its velocity. Find an expression for the time it takes the projectile to reach its highest point. For $m = 100$ g and $v_0 = 20$ m/s, plot this time versus the drag coefficient b for $0 \leq b \leq 1$. Confirm your value for t when $b = 0$.*

Let y measure vertical height, so $v = \dot{y}$. Newton's Second Law says

$$ma = m\dot{v} = F = -mg - bv$$

so solve that (first order) differential equation subject to the initial condition $v(0) = v_0$, and find the time at which $v = 0$.

Notebook 4.1 show a solution to this problem. Solving the differential equation just follows what we outlined in Sec. 4.1. I didn't bother to print out the form of the solution for $v(t)$, but I used that solution to define an expresseion "vel" which in turn is set equal to zero and solved for the time.

As in Example 2.3, I specified the domain "Reals" when using "Solve", to avoid the complex solutions possible when dealing with logarithms. The solution looks odd at first, but you'll see physics in there when you look at it more closely. It is a "ConditionalExpression" that depends on the sign of b, g, m, and of the expression $gm/b + v_0$.

We set up the problem assuming that b, g, and m are all positive numbers,

NOTEBOOK 4.1 Solution to Example 4.1

```
In[1]:= Remove["Global`*"]
```

A Vertical Projectile with Drag

Solve the differential equation

```
In[2]:= sol = DSolve[{m v'[t] == -m g - b v[t], v[0] == v0}, v, t];
        solAlone = Part[Part[sol, 1], 1];
        vel = v[t] /. solAlone;
```

```
In[5]:= solt = Solve[vel == 0, t, Reals]
```

Out[5]=

$$\left\{\left\{t \to \text{ConditionalExpression}\left[\frac{m \,\text{Log}\left[\frac{g\,m+b\,v0}{g\,m}\right]}{b},\right.\right.\right.$$
$$\left(b>0\,\&\&\,g>0\,\&\&\,m>0\,\&\&\,\frac{g\,m}{b}+v0>0\right) \,||\, \left(b>0\,\&\&\,g>0\,\&\&\,m<0\,\&\&\,\frac{g\,m}{b}+v0<0\right) \,||$$
$$\left(b>0\,\&\&\,g<0\,\&\&\,m>0\,\&\&\,\frac{g\,m}{b}+v0<0\right) \,||\, \left(b>0\,\&\&\,g<0\,\&\&\,m<0\,\&\&\,\frac{g\,m}{b}+v0>0\right) \,||$$
$$\left(b<0\,\&\&\,g>0\,\&\&\,m>0\,\&\&\,\frac{g\,m}{b}+v0<0\right) \,||\, \left(b<0\,\&\&\,g>0\,\&\&\,m<0\,\&\&\,\frac{g\,m}{b}+v0>0\right) \,||$$
$$\left.\left.\left.\left(b<0\,\&\&\,g<0\,\&\&\,m>0\,\&\&\,\frac{g\,m}{b}+v0>0\right) \,||\, \left(b<0\,\&\&\,g<0\,\&\&\,m<0\,\&\&\,\frac{g\,m}{b}+v0<0\right)\right]\right\}\right\}$$

Make the plot with the specified values

```
In[6]:= vals = {m -> 0.1, g -> 9.8, v0 -> 20};
        Plot[t /. solt /. vals, {b, 0, 1},
         PlotRange -> {{0, 0.5}, {0, 2.5}}]
```

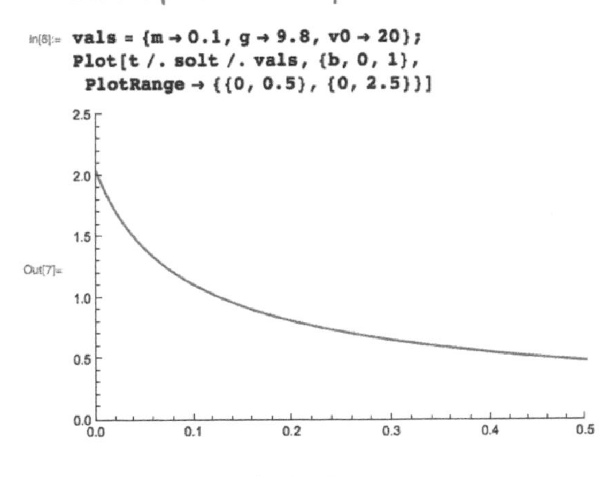

Check the result for b=0

```
In[8]:= v0 / g /. vals

Out[8]= 2.04082
```

but MATHEMATICA isn't aware of that, of course. Nevertheless, when we evaluate the expression, we expect all these to be true. You'll see how this works in a moment when we make the plot of time versus b. On the other hand, if you really want to have a "clean" expression without all the conditions, you can execute something like

```
time = t /. solt;
Simplify[time,
   Assumptions -> {b > 0 && g > 0 && m > 0 && g m + b v0 > 0}]
```

and the expression "time" will have none of the embedded conditions.

Consider now the physics contained in the sign of $gm/b + v_0$. This will be positive because each variable is positive. This includes v_0, since we are lauching m upwards, but suppose b were negative. This is entirely unphysical, a "negative drag" force that causes the mass to accelerate. Nevertheless, if $b < 0$ and $gm/b + v_0 < 0$, then $|b|v_0 > mg$ and at launching, there is a vertical force that overpowers gravity, and just gets larger as the mass accelerates. Indeed, the mass would never come to rest.

Now let's return to analyzing the solution in Notebook 4.1. We define a list "vals" that contains the values specified in the problem. We then plot "t" implementing its replacement "solt" and then replacing m, g, and v_0 with the specified values. The range of values to plot for b is also given by the specified values, and we use the "Plot" option to specify the plot range for b and the vertical range for the time. I chose a different plot range for b than the range over which we asked for the plot, just to show you that it can be done. The vertical range was chosen to make sure there was no suppressed zero on the vertical axis.

If there was no drag, that is $b = 0$, then $v = v_0 - gt$ and the mass would come to a stop at $t = v_0/g$. This is the value calculated, with the same replacement "vals", and the number 2.04082 does appear to agree with the vertical intercept on the plot.

Example 4.2 *A mass $m = 100$ g, initially at rest, is dropped from a height $h = 100$ m and is subject to a linear drag force bv, proportional to its velocity. Find an expression for its height as a function of time, and plot it until it hits the ground for various values of b. Comment on the shape of the curves.*

The differential equation and initial conditions we need to solve are

$$\begin{aligned} m\ddot{y} &= -mg - b\dot{y} \\ y(0) &= h \\ \dot{y}(0) &= 0 \end{aligned} \tag{4.2}$$

and the rest we know how to do in MATHEMATICA. I will embellish the solution, though, to introduce a couple of new things along the way, to better understand the physics.

NOTEBOOK 4.2 Solution to Example 4.2

```
In[1]:= Remove["Global`*"]
```

A Falling Ball with Drag

Solve the differential equation

```
In[2]:= DSolve[{m y''[t] == -m g - b y'[t], y[0] == h, y'[0] == 0}, y, t];
        height = Simplify[Part[y[t] /. %, 1]]
```

Out[3]= $$\frac{b^2 h + \left(1 - e^{-\frac{b t}{m}}\right) g m^2 - b g m t}{b^2}$$

Check the limit as b goes to zero

```
In[4]:= Limit[height, b → 0]
```

Out[4]= $$h - \frac{g t^2}{2}$$

Use values to plot

```
In[5]:= mgh = {m → 0.1, g → 9.8, h → 100};
        Manipulate[
         Plot[height /. mgh /. b → drag, {t, 0, 15},
          PlotRange → {0, 100}],
         {drag, 0.0001, 1}]
```

Out[6]=

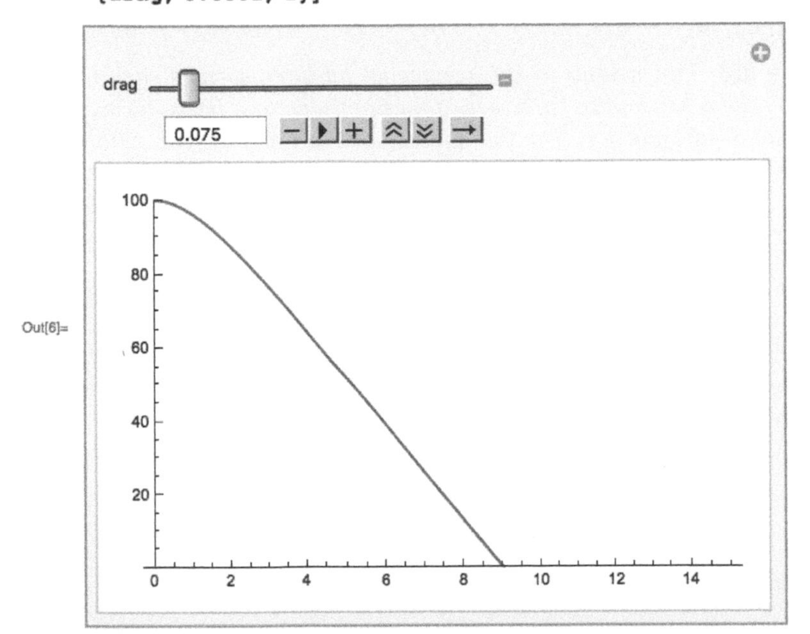

See Notebook 4.2 for my solution. The differential equation is solved in the standard way, and I turn the result into an expression called "height" after extracting the list element and using "Simplify" to get a cleaner expression. (You might try experimenting and see what the answer looks like if you don't simplify.) Note that I was lazy, and instead of defining the solution as another variable, I just used "%" to define "height".

It's always good to check your answer in one way or another. In this case, we know that for $b = 0$, we expect $y = h - gt^2/2$. However, we get an indeterminate expression if we just try to evaluate "height" by setting $b = 0$. (You should try it and see.) We can, on the other hand, use the MATHEMATICA function "Limit" to check to see if you get this answer for $b \to 0$, and indeed it works out.

In order to make a plot for "various values of b", I'm introducing a new MATHEMATICA function, namely "Manipulate". A "Plot" command, using specified values for the other parameters, is imbedded inside "Manipulate". The value of b is replaced with a variable "drag", and this variable is used in "Manipulate". This allows you to use the slider on the plot to set b to different values, and the plot changes dynamically.

Of course, you should try this out and see for yourself how "Manipulate" works. The plot shown is for $b = 0.075$, which is small enough to see the parabolic behavior at early times, when the velocity is small and the drag force is approximately negligible. However, at larger times, the height falls linearly with time because the mass is falling at the (constant) terminal velocity.

I should warn you, though, about "Manipulate" when using expressions defined outside the command. We will discuss this in Section 11.2, but the problem has to do with the concept of *scoping* of variables. For now, though, just try to stick with manipulating plots of expressions entered directly into the "Plot" command, and you should be alright.

Example 4.3 *Two masses m are connected by three springs each with stiffness k to each other and to fixed walls on either side of a frictionless, horizontal surface. See Figure 4.1. The two masses are initially at rest, with the mass on the left displaced by a distance A and the other mass at its equilibrium position. Find and plot the positions of each mass as a function of time, in units of $\tau \equiv 2\pi(m/k)^{1/2}$. Also plot the sum and difference of their positions.*

Label the position of the mass on the left (right) by x_1 (x_2), where $x_{1,2} = 0$ marks the equilibrium positions. For positive (negative) extension of the mass on the left (right) the force from the outer springs will be towards the left. If the spring in the middle is stretched ($x_2 - x_1 > 0$) then the force from this spring on the left (right) mass will be towards the right (left). This leads us to write the coupled equations of motion as

$$m\ddot{x}_1 = -kx_1 + k(x_2 - x_1) \quad (4.3a)$$

$$m\ddot{x}_2 = -kx_2 - k(x_2 - x_1) \quad (4.3b)$$

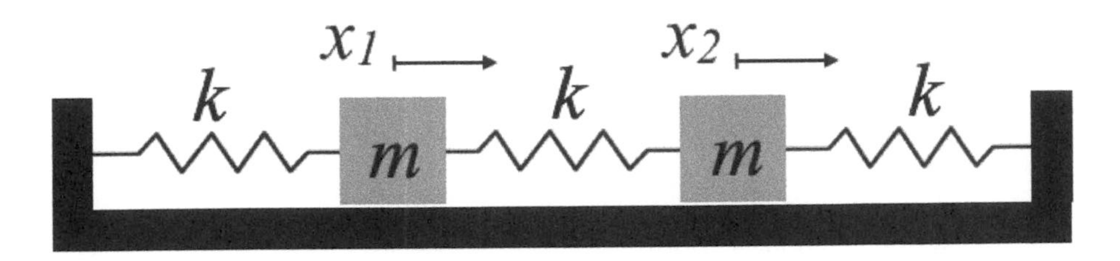

FIGURE 4.1 Diagram to go with Example 4.3

From here, it is just a straightforward application of MATHEMATICA commands that we have learned in this and previous chapters.

Notebook 4.3 shows one way to approach the solution and the rest of the problem, along with some embellishments. We'll go through this step by step.

First we define the differential equations and solve them, exactly analogous to what we did in Section 4.3. The two equations and four initial conditions are given to "DSolve", and we ask it to return the solutions for $x_1(t)$ and $x_2(t)$. We then extract the solutions from the list returned by "DSolve", and store them in the expressions "x1sol" and "x2sol". From here, it's just a matter of using what we've learned in earlier chapters.

The problem asks us to scale the time variable by $\tau \equiv 2\pi(m/k)^{1/2}$, which of course is just the period of oscillation for a single mass m attached to a single spring with stiffness k. We define the variable τ in the usual way[2] and make the substitution $t = t_{\text{Scale}}\tau$. Notice, however, that we do not simply make the substitution and be done with it. Indeed, you should see what you get if you simply enter "x1plot=x1sol/.subs" at this point. You will not find the expected cancellation of dependence on m and k. You can try using "Simplify", but that won't fix the problem.

The reason is that expressions like $\sqrt{m}\sqrt{k/m}/\sqrt{k}$ do not simplify on their own, because if m or k are complex numbers, the square root may not have well defined values. So, before using "Simplify", we use the concept of *Global Assumptions* to make it known that from here on, m and k are positive real numbers. (Instead of making global assumptions, we could also use the "Assuming" option inside "Simplify". See Page 50.) Then, all the cancellations work like we expect them to.

I suppressed the output in the solution only to save space in the figure. If you remove the semicolon and look at the output, you'll find that the solutions (with $t_{\text{Scale}} = t/\tau$ replaced here by t)

$$x_1(t) = \frac{5}{4}\left(e^{-2i\pi t} + e^{2i\pi t} + e^{-2i\sqrt{3}\pi t} + e^{2i\sqrt{3}\pi t}\right) \tag{4.4a}$$

$$x_2(t) = \frac{5}{4}\left(e^{-2i\pi t} + e^{2i\pi t} - e^{-2i\sqrt{3}\pi t} - e^{2i\sqrt{3}\pi t}\right) \tag{4.4b}$$

[2]Recall how to enter non-Latin characters, for example using LaTeX and the escape key.

NOTEBOOK 4.3 Solution to Example 4.3

```
In[1]:= Remove["Global`*"]
```

Coupled mass and spring oscillations

Set up and solve the coupled equations

```
In[2]:= eq1 = m x1''[t] == -k x1[t] + k (x2[t] - x1[t]);
        eq2 = m x2''[t] == -k x2[t] - k (x2[t] - x1[t]);
        sol = DSolve[{eq1, eq2,
              x1[0] == A, x1'[0] == 0, x2[0] == 0, x2'[0] == 0},
            {x1, x2}, t];
```

Extract the solutions

```
In[5]:= x1sol = x1[t] /. Part[Part[sol, 1], 1];
        x2sol = x2[t] /. Part[Part[sol, 1], 2];
```

Massage the solutions so they can be scaled

```
In[7]:= τ = 2 Pi Sqrt[m / k];
        subs = {A → 5, t → tScale τ};
        $Assumptions = m > 0 && k > 0;
        x1plot = Simplify[x1sol /. subs];
        x2plot = Simplify[x2sol /. subs];

In[12]:= ExpToTrig[x1plot];
         ExpToTrig[x2plot];
```

Make the plots

```
In[14]:= Plot[{x1plot, x2plot}, {tScale, 0, 2},
          PlotStyle → {Solid, Dashed}, PlotLegends → {x1, x2}]
         Plot[{x1plot + x2plot, x1plot - x2plot}, {tScale, 0, 2},
          PlotStyle → {Solid, Dashed}, PlotLegends → {x1 + x2, x1 - x2}]
```

These expressions in terms of exponentials are perfectly natural ways to write the solution. They aren't necessarily how you'd write them to get some physical insight into the solution, however.

For this reason, I added the command "ExpToTrig" in the notebook to manipulate the expressions. Again, the output is suppressed for reasons of space, but if you look at the result, you'll find

$$x_1(t) = \frac{5}{2}\cos(2\pi t) + \frac{5}{2}\cos\left(2\sqrt{3}\pi t\right) \tag{4.5a}$$

$$x_2(t) = \frac{5}{2}\cos(2\pi t) - \frac{5}{2}\cos\left(2\sqrt{3}\pi t\right) \tag{4.5b}$$

The "Plot" command for $x_1(t)$ and $x_2(t)$ results in the following:

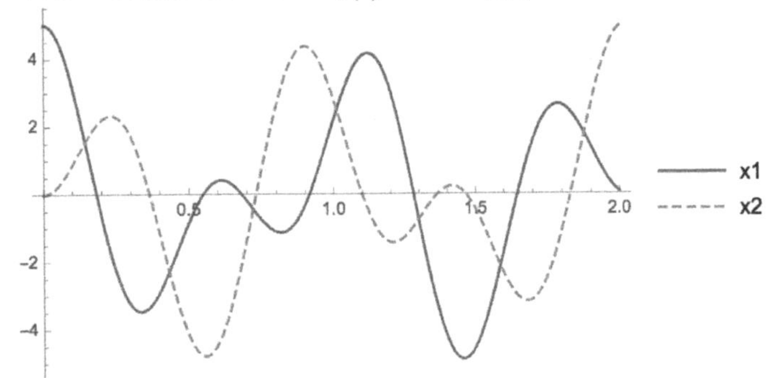

The oscillatory behavior is a bit peculiar, apparently periodic, but obviously not purely sinusoidal, for either mass. That is not unexpected, since this is not a simple "one mass, one spring" system. Instead, the masses go through some coupled motion where the first moves towards equilibrium after being released, with the changing force on the second mass causing it to move off equilibrium.

On the other hand, the plot of the sum and difference gives us some insight:

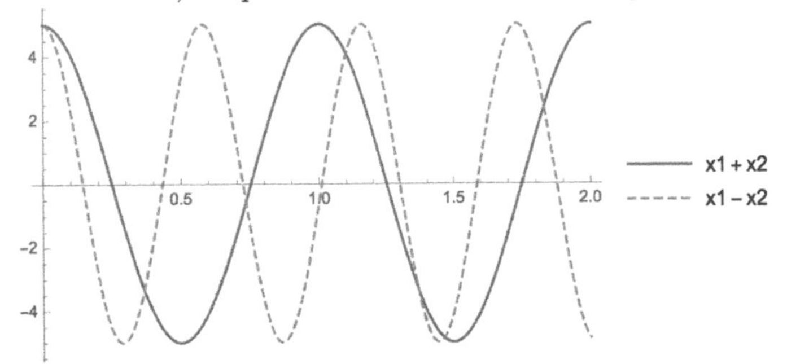

These *do* appear to be sinusoidal, one having frequency $\sqrt{k/m}$ and the other somewhat faster. Indeed, from the form of the solution given in (4.5) we see that they are indeed sinusoidal, with the difference having frequency $\sqrt{3k/m}$.

If you have taken an upper level course in classical mechanics, you will recognize this as an example of an eigenvalue problem in coupled oscillations. We will return to this when we cover matrix manipulations in Chapter 6.

4.6 CHAPTER SUMMARY

- Use "DSolve" to solve differential equations in pretty much the same way that you use "Solve" for algebraic equations.
- Specify initial/boundary conditions as additional equations in a list, otherwise a solution is returned in terms of integration constants.
- Simultaneous or second (or higher) order differential equations are solved the same way, but they require more boundary conditions or return additional integration constants.
- Partial differential equations are handled similarly, but be aware that the uniqueness of the solution may depend on the boundary conditions.

EXERCISES

4.1 *The damped harmonic oscillator is a mass m that responds to both a restoring force $-kx$ and a damping force $-bv$, where $x(t)$ and $v(t) = dx/dt$ are the position and velocity of the mass, and b and k are positive constants. The motion is strikingly different depending on whether $b^2 < 4km$ ("Under Damped"), $b^2 > 4km$ ("Over Damped"), or $b^2 = 4km$ ("Critically Damped"). Solve the differential equation of motion for each of these cases, assuming the oscillator starts from rest someplace away from equilibrium, and plot the motion for the three cases. Use the same values for k and m for the three cases, and three different values for b.*

4.2 *A circuit consists of a battery V, switch S, resistor R, and capacitor C connected in series, as shown in the figure below.*

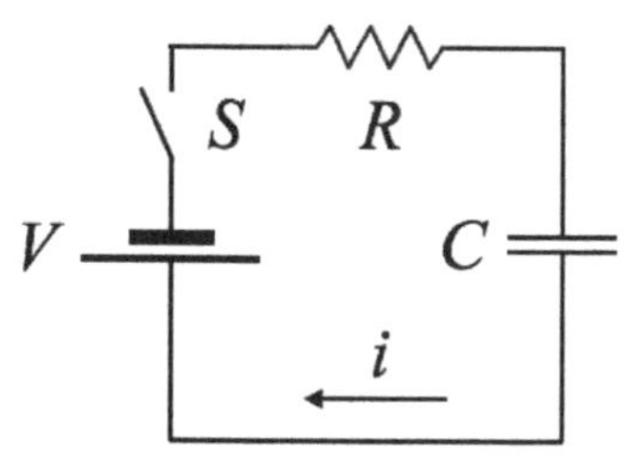

Recall that the voltage drop on the capacitor holding a charge q is q/C and the current $i = dq/dt$. Find $i(t)$ assuming that the capacitor is uncharged when the switch is closed at $t = 0$.

4.3 *A circuit consists of a battery V, switch S, resistor R, and inductor L connected in series, as shown in the figure below.*

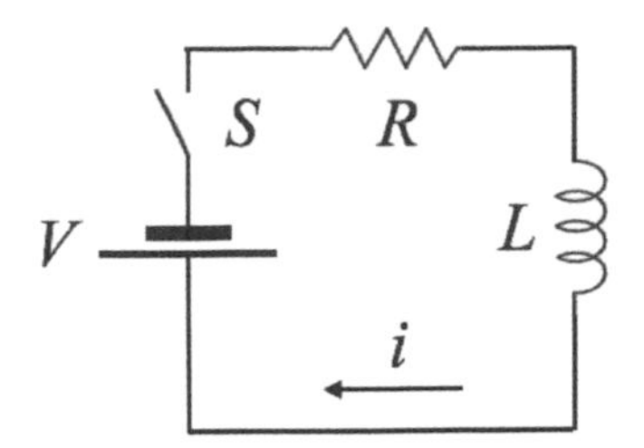

Recall that the voltage drop on the inductor carrying a current i is Ldi/dt. Find $i(t)$ assuming that no current is flowing when the switch is closed at $t = 0$.

4.4 *A circuit consists of a battery V, switch S, resistor R_1, and capacitor C connected in series, as shown in the figure below.*

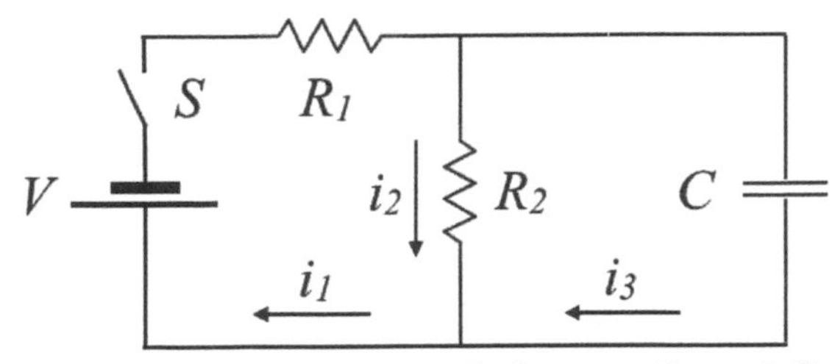

A second resistor R_2 is connected to a node between R_1 and C and in parallel to C. The switch is initially open with no currents flowing and the capacitor is uncharged. Find the currents $i_1(t)$, $i_2(t)$, and $i_3(t)$ as a function of time, in terms of V, R_1, R_2, and C . Using $V = 10$ V, $R_1 = 10$ $k\Omega$, $R_2 = 20$ $k\Omega$, and $C = 500$ μF, plot the three currents (in mA) as a function of time, and argue that your result appears to give the correct behavior.

4.5 *Two masses are connected to two springs as shown in the figure below.*

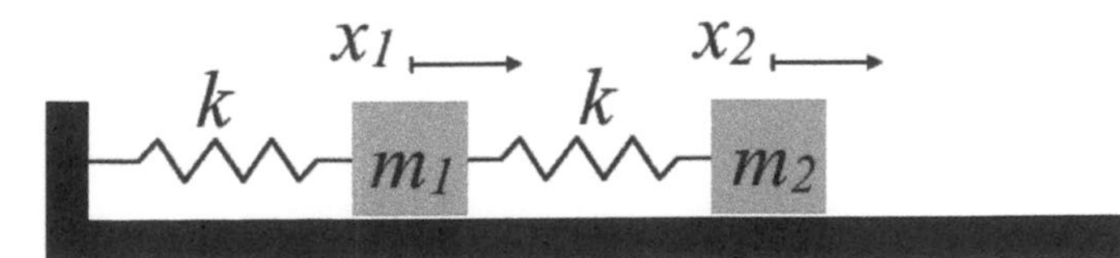

The masses move horizontally without friction, and can have different values. The springs are identical. Write the differential equations of motion and cast them into a form that depends only on the quantities ω_1 and ω_2 where $\omega_1^2 = k/m_1$ and $\omega_2^2 = k/m_2$. Find a general solution to the equations of motion, where both masses start from rest, m_1 starts from equilibrium, and m_2 starts with displacement A. Plot the motion for $\omega_1 = \omega_2$, $\omega_1 \ll \omega_2$, and $\omega_1 \gg \omega_2$.

CHAPTER 5

Differential Equations: Numerical Solutions

CONTENTS

Physicists often encounter differential equations that cannot be solved analytically, even in terms of special functions. In this case, we have no choice but to solve them numerically, and MATHEMATICA makes this a very painless procedure.

Of course, there are many algorithms that might be used for numerical solutions of differential equations. Beginners, however, can rely on MATHEMATICA to make the best choice for a solution under typical circumstances. The documentation will give you more information and more options.

Numerical solution is possible for the same classes of problems for which we've discussed analytic solutions, namely ordinary and partial differential equations and systems of equations. This chapter will briefly discuss the syntax in MATHEMATICA and then move on to a few examples, followed by exercises.

Physicists should note a special point when using numerical solutions to differential equations. You always want to keep fixed parameters (such as mass, gravitational constants, resistances, ...) symbolic in your solutions, but you cannot do that when solving equations numerically. Rather than just setting these parameters to values, however, you should consider *scaling* your equations so that they are dimensionless. This way, you can always return to the general case through the scale parameters. See Section 5.4 for examples.

5.1 ORDINARY DIFFERENTIAL EQUATIONS

Suppose we want to solve the differential equation

$$\frac{d^2y}{dx^2} = y^2 - 2x$$

subject to the initial conditions $y(0) = 0$ and $y'(0) = 1$. We execute the following cell in MATHEMATICA:

```
ode = D[D[f[x], x], x] == f[x]^2 - 2 x ;
ics = {f[0] == 0, f'[0] == 1};
DSolve[{ode, ics}, f[x], x]
```

However, MATHEMATICA just returns the same thing we gave it. It is unable to find an analytic solution to this differential equation. Indeed, such a solution does not exist.

The MATHEMATICA function “NDSolve” finds *numerical* solutions to differential equations. Of course, there can be no symbolic parameters in the equation, and we have to tell MATHEMATICA the range over which we want to find a solution, but that's simple to do. For the problem above, we might try something like the following:

```
ode = D[D[f[x], x], x] == f[x]^2 - 2 x ;
ics = {f[0] == 0, f'[0] == 1};
xmax = 10;
sol = NDSolve[{ode, ics}, f[x], {x, 0, xmax}];
y = f[x] /. Part[Part[sol, 1], 1];
Plot[y, {x, 0, xmax}]
```

(Note that I defined a value for “xmax”, and used this parameter both in the solution and in the plot. This is handy for making sure I don't try to plot outside the limits of the solution.) The plot looks like this:

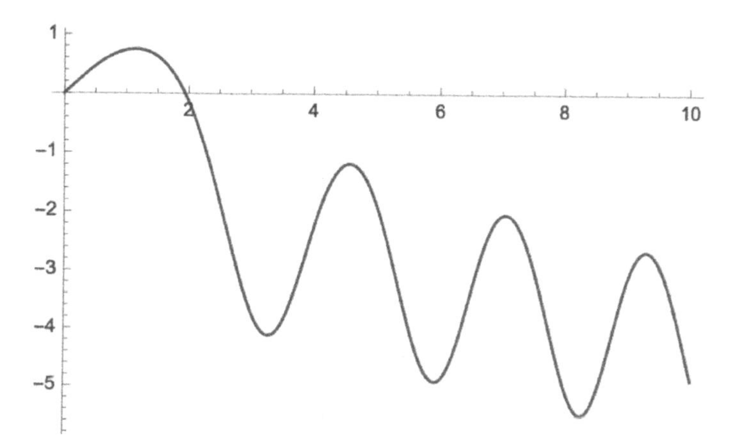

This result looks right. The initial conditions are satisfied, and the second derivative starts out negative but then changes sign as the y^2 term competes with the $-2x$ term.

If you ask MATHEMATICA to give you the expression for "y", you will be told that is an "InterpolatingFunction" with a peculiar-looking argument. (In the latest versions of MATHEMATICA you will actually see a little picture of the plot.) You can read the documentation if you want details, but you get the idea. Over the range for which the solution was found, MATHEMATICA will interpolate through the points you give it and reproduce the plot. You can ask for values as well. For example, "y/.x → 4" returns the value "-2.25506".

5.2 PARTIAL DIFFERENTIAL EQUATIONS

You already have the machinery for solutions, numerically or otherwise, for solving partial differential equations. So, let's do an example that can serve as sort of a review and then get into some extensions.

Consider the partial differential equation

$$\frac{\partial^2 u}{\partial x^2} + \frac{\partial^2 u}{\partial y^2} = 0 \tag{5.1}$$

Depending on how far you are in your physics courses, you may recognize this as *Laplace's Equation*, which is typically encountered in an upper level course in electromagnetism. In that case, $u(x, y)$ is the static electric potential in two dimensions.

This equation can be solved analytically in MATHEMATICA. The cell

```
pde = D[D[u[x, y], x], x] + D[D[u[x, y], y], y] == 0;
DSolve[pde, u[x, y], x, y]
```

returns the result

```
{{u[x, y] -> C[1][I x + y] + C[2][-I x + y]}}
```

In other words, the solution to Equation 5.1 is

$$u(x, y) = f(ix + y) + g(-ix + y) \tag{5.2}$$

where $f(\xi)$ and $g(\eta)$ are arbitrary functions of one variable. It's not hard to see that this is the correct solution. Differentiating by x twice gives you the same result as differentiating twice by y, but with an extra factor of $(\pm i)^2 = -1$.

Now suppose you want to add boundary conditions. For an electrostatics problem, you might let $u(x, y)$ represent the electric potential in a region bounded by infinitely long conducting plates (parallel to the z-axis) that form a square in two dimensions. If three of the plates are grounded and the fourth is held at a fixed, constant potential (say, 10V), then the boundary conditions might be

$$u(-1, y) = u(1, y) = u(x, -1) = 0 \qquad \text{and} \qquad u(x, 1) = 10 \tag{5.3}$$

You would then use something like the following cell to find a solution:

```
pde = D[D[u[x, y], x], x] + D[D[u[x, y], y], y] == 0;
bcs = {u[-1,y] == 0, u[1,y] == 0, u[x,-1] == 0, u[x,1] == 10};
DSolve[{pde, bcs}, u[x, y], x, y]
```

However, MATHEMATICA just returns what you typed in. It cannot find an analytic solution to the equation with these boundary conditions imposed.

Think for a moment about what this means. Equation 5.1 is a linear differential equation, so we can have any number of solutions of the form 5.2 and the result is still a solution. In addition, Equation 5.2 is a very general form, so it seems like we should have the ability to form a solution that meets the boundary conditions 5.3.

For example, suppose $f(\xi) = g(\xi) = e^{\pm ik\xi}$. We could then form a solution by summing (or integrating) over values of k that would involve sines, cosines, etc..., of x and y. With appropriate choices of the coefficients in that sum, we might expect to meet the boundary conditions. This is likely an infinite sum (or integral), however, and not a closed analytic form.

Many of you will recognize where we are going with this, namely forming Fourier series or Fourier transforms. You can of course do this mathematics in MATHEMATICA, but we won't be covering these topics in this textbook. Nevertheless, you should now have a feel for how "DSolve" can have too much trouble with a partial differential equation with boundary conditions.

Now return to the problem of solving 5.1 with the boundary conditions 5.3, but instead attempt a numerical solution. The cell might look like

```
pde = D[D[u[x, y], x], x] + D[D[u[x, y], y], y] == 0;
bcs = {u[-1,y] == 0, u[1,y] == 0, u[x,-1] == 0, u[x,1] == 10};
sol = NDSolve[{pde, bcs}, u[x, y], {x, -1, 1}, {y, -1, 1}];
z = u[x, y] /. Part[Part[sol, 1], 1];
```

Compared to our attempt at an analytic solution, I have replaced "DSolve" with "NDSolve", and included a range of values for x and y that coincide with our boundary conditions. I have also extracted the solution and stored it in an expression $z(x, y)$.

We should look at the result and see if it makes sense. The following command makes a plot of $u(x, y)$ as a function of y over the full range $-1 \leq y \leq 1$, for fixed values of x along the mid-plane ($x = 0$) and several values of x getting closer and closer to one edge:

```
Plot[
 {z /. x -> 0, z /. x -> 0.5, z /. x -> 0.75, z /. x -> 0.95},
 {y, -1, 1},
 PlotStyle -> {Solid, Dashed, Dotted, DotDashed},
 PlotLegends -> {0, 0.5, 0.75, 0.95}]
```

The output of this plot is shown in Figure 5.1. The boundary conditions $u(-1, y) = 0$ and $u(1, y) = 1$ are clearly satisfied for all four values of y. It is

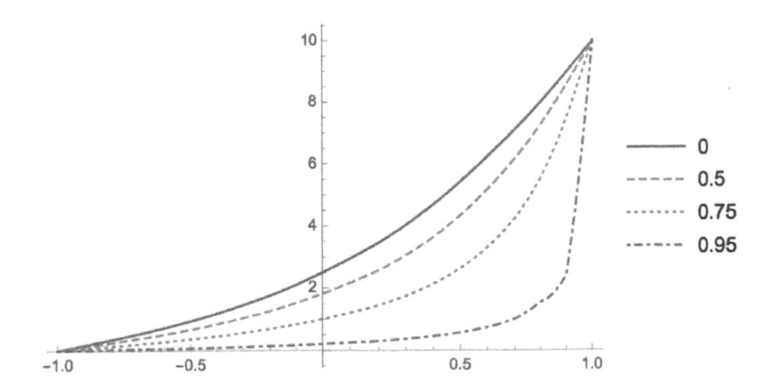

FIGURE 5.1 A two-dimensional representation of the solution to the problem in Section 5.2.

also clear that as x gets closer to the edge, u gets closer to zero over most of the range, so that $u(1, y) = 0$ is also satisfied.

Notice that the curve for $u(0.95, y)$ is not quite as smooth as the other curves, particularly in the region close to $y = 1$. This is a difficult region to numerically calculate accurately, because of the discontinuity at $y = 1$. I've only used the default algorithms in MATHEMATICA in this example, but there are other algorithms you can use, and parameters you can adjust, to increase the accuracy of certain calculations. See the Documentation for more details.

5.3 PLOTTING IN THREE DIMENSIONS

Let's take a short detour and take advantage of the solution in Section 5.2 to introduce options for plotting in three dimensions.

Figure 5.2 shows three different three-dimensional plots of the solution $z(x, y)$ to Laplaces' equation, where $z(-1, y) = z(1, y) = z(x, -1) = 0$ and $z(x, 1) = 1$, as a function of x and y. In order from left to right, these plots were made using the MATHEMATICA commands "Plot3D", "DensityPlot", and "ContourPlot", sometimes making use of optional arguments. Let's go through them one by one.

The command

```
Plot3D[z, {x, -1, 1}, {y, -1, 1}]
```

produced the first plot. The result is simply a rendering of the surface that has the shape $z(x, y)$, including labeled tick marks on the x, y, and z axes. The grid lines and shadowing help you see the shape of the surface. In fact, you should be able to trace the shape of the surface along lines of constant x to see that they agree with the lines in Figure 5.1.

The second plot was made using

```
DensityPlot[z, {x, -1, 1}, {y, -1, 1}]
```

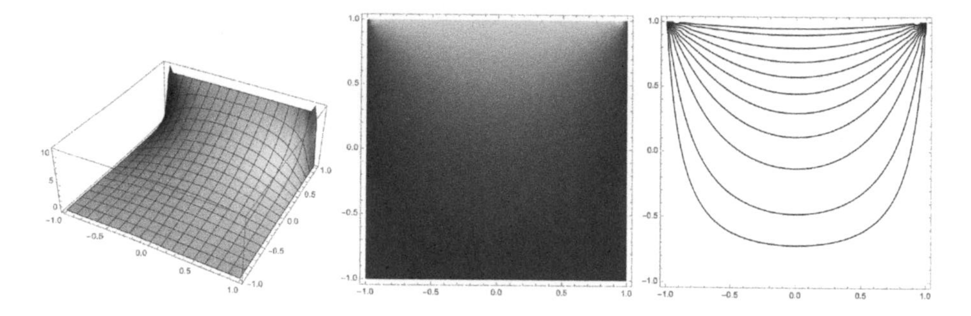

FIGURE 5.2 Three different three-dimensional representations of the solution to the problem in Section 5.2. From left to right, they are produced by "Plot3D", "DensityPlot", and "ContourPlot". See the text for details.

The shading intensity is proportional to the magnitude of z, and you can generally see the behavior of the function. This kind of plot is not ideal for quantitative information, but it does a good job conveying the general characteristics of $z(x, y)$ for a wider audience.

Perhaps the best way to get across quantitative information is using a contour plot, such as the third in Figure 5.2, made using the command

```
ContourPlot[z, {x, -1, 1}, {y, -1, 1},
 ContourShading -> None,
 Contours -> {0.5, 1, 2, 3, 4, 5, 6, 7, 8, 9, 9.5}]
```

In this case, I made use of some options. The default is to shade the plots (to help you interpolate in between contours) but I turned that off. I also specified the values of the contours; they are not labeled in the plot (which is another option) but it should be clear that the lowest contour corresponds to $z = 0.5$ and that the one closest to the top of the figure is for $z = 9.5$.

There are a few other three-dimensional plotting commands, and plenty of options for all of them, so you are invited to explore these by following up with the Documentation.

5.4 PHYSICS EXAMPLES

Example 5.1 *An object of mass m moves in one dimension x under a force $F(x) = ax^2 - bx$, where a and b are positive constants. Solve for $x(t)$ with initial conditions $\dot{x}(0) = 0$ and $x(0) = x_0$, and plot the results, for $0 \leq t \leq t_{\text{Max}}$ where t_{Max} is large enough for you to see the behavior as $t \to \infty$. Let x_0 take on each of three values, with (a) $|x_0| \ll b/2a$, (b) $x_0 > -b/2a$ (by a small amount), and (c) $x_0 < -b/2a$ (also by a small amount). Find a physical reason to explain why $x = -b/2a$ is special.*

The differential equation of motion is

$$m\frac{d^2x}{dt^2} = ax^2 - bx \tag{5.4}$$

Let's first convert this equation to dimensionless form. One obvious length scale is $\alpha = b/2a$ so a dimensionless length is $y \equiv x/\alpha$. This leads to

$$m\frac{d^2y}{dt^2} = a\alpha y^2 - by = \frac{b}{2}y^2 - by$$

A natural time scale is $\beta = \sqrt{2m/b}$. (Check these dimensions for yourself!) So, a dimensionless time is $u \equiv t/\beta$, and we get

$$\frac{d^2y}{du^2} = y^2 - 2y \tag{5.5}$$

This equation is dimensionless. The initial conditions are $\dot{y}(0) = 0$ and $y(0) = y_0 \equiv x_0/\alpha$ where (a) $|y_0| \ll 1$, (b) $y_0 = -1 + \varepsilon$ ($\varepsilon \ll 1$), and (c) $y_0 = -1 - \varepsilon$. After having a solution, we can always convert back to x and t from y and u by multiplying by α and β. If we want energy, the scale is $m\alpha/\beta^2 = b^2/4a$.

To be sure, this differential equation can be solved with "DSolve", but the result is in terms of special functions and algebraic roots. These need to be evaluated numerically anyway, so we might as well just use "NDSolve" and work numerically from the start.

Notebook 5.1 shows one solution to this problem. Let's go through it step by step.

Firstly, as is my habit, the differential equation solver comes up with a function "f[u]", which we later turn into an expression. In the first cell, we define the differential equation, choose values of y_0 for the initial starting point, according to the problem specifications, and then set up the initial conditions as lists. The first cell also defines our choice for the maximum value of (dimensionless) time.

The next cell uses "NDSolve" to numerically find solutions for the three different sets of initial conditions. Expressions are set up for each solution, namely "ya", "yb", and "yc", stripping out the solution from the nested lists. (As you likely guessed, "NDSolve" returns solutions in the same way as "DSolve" and "Solve".)

Then, each of the cases are plotted. The plot output is suppressed in Notebook 5.1 to save space, but the three plots (a), (b), and (c) are shown here:

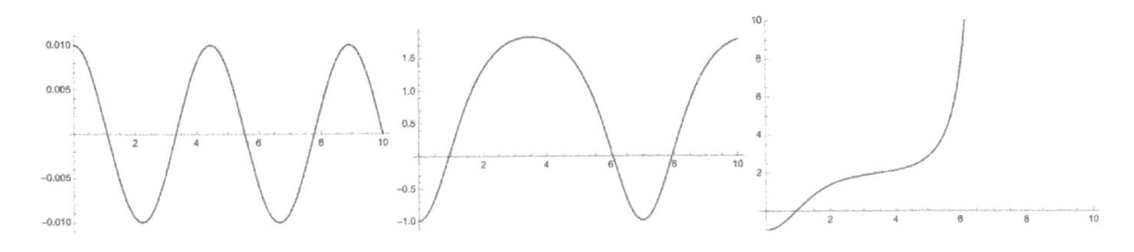

NOTEBOOK 5.1 Solution to Example 5.1.

```
In[1]:= Remove["Global`*"]
```

A Humpy Potential Well

Set up the equation, initial conditions, and max time

```
In[2]:= diffeq = D[D[f[u], u], u] == f[u]^2 - 2 f[u];
        y0a = 0.01; y0b = -0.99; y0c = -1.01;
        ica = {f'[0] == 0, f[0] == y0a};
        icb = {f'[0] == 0, f[0] == y0b};
        icc = {f'[0] == 0, f[0] == y0c};
        umax = 10;
```

Solve the equation for the three initial conditions

```
In[8]:= sol = NDSolve[{diffeq, ica}, f[u], {u, 0, umax}];
        ya = f[u] /. Part[Part[sol, 1], 1];
        sol = NDSolve[{diffeq, icb}, f[u], {u, 0, umax}];
        yb = f[u] /. Part[Part[sol, 1], 1];
        sol = NDSolve[{diffeq, icc}, f[u], {u, 0, umax}];
        yc = f[u] /. Part[Part[sol, 1], 1];
```

Plot the three cases

```
In[12]:= Plot[ya, {u, 0, umax}]
```

```
In[13]:= Plot[yb, {u, 0, umax}]
```

```
In[14]:= Plot[yc, {u, 0, umax}, PlotRange → {-1, 10}]
```

Consider the "potential energy"

```
In[15]:= poten = -Integrate[ξ^2 - 2 ξ, {ξ, 0, y}];
         Plot[poten, {y, -1.1, 3.5}]
```

The motion is indeed rather different for all three cases. When the mass starts out near $x = 0$, it oscillates in what looks to be very close to a cosine function, that is, simple harmonic motion. If the starting point is just to the right of $x = -b/2a$, the mass still oscillates, but it is clearly no longer simple harmonic motion.

On the other hand, if it starts out just to the left of $x = -b/2a$, it moves to the right and slows down almost to a stop, but then picks up speed rapidly. In fact, you'll see that I added the option "PlotRange→{-1,10}" to the "Plot" command in Notebook 5.1, so that figure would stay in a reasonable range.

If you consider the (scaled) potential energy $U(y) = -\int_0^y (\xi^2 - 2\xi) d\xi$, then the explanation for this weird behavior is clear. The notebook calculates and plots the potential energy curve, but I have once again suppressed the plot:

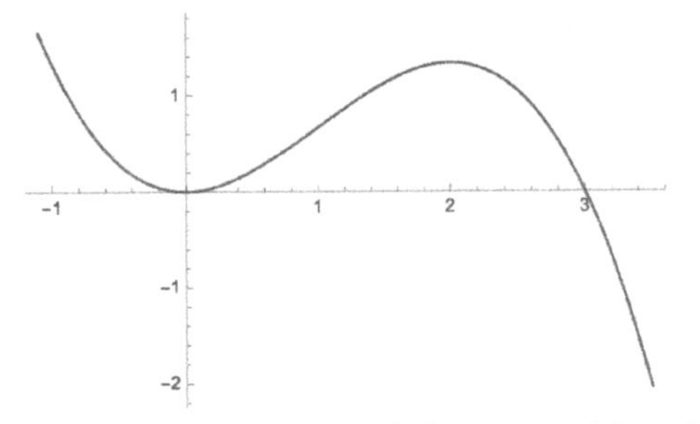

Now you see that $x = 0$ is the bottom of the potential well, and oscillations in that vicinity will be (approximately) simple harmonic. Asymmetric oscillations will occur for larger excursions from equilibrium, that is, case (b).

However, if you start anywhere in the region $x < -b/2a$, i.e. $y < 1$, then the mass has enough energy to go over the top of the "hump" at $y = 2$, i.e. $x = b/a$, and then rapidly move off to $+\infty$. This is exactly the behavior given by the MATHEMATICA solution.

Example 5.2 *A long straight thin metal rod has length ℓ. Its temperature $T = T(x, t)$ at time t, where x measures position along the rod, is governed by*

$$\frac{\partial T}{\partial t} = \kappa \frac{\partial^2 T}{\partial x^2} \tag{5.6}$$

and the thermal diffusivity constant κ is a property of the metal. The rod starts out at a uniform temperature $T(x, 0) = T_0$, but with one end in contact with an ice batch at $T = 0°\,C$ and the other in boiling water at $T = 100°\,C$. Determine and plot $T(x, t)$ for different values of T_0. Show the temperature distribution both for "short" and "long" times. (What are the natural length and time scales?)

This is a classic example in the mathematics of heat transport, and the problem can be solved analytically in terms of Fourier series, but we will take the numerical approach here.

NOTEBOOK 5.2 Solution to Example 5.2.

```
In[1]:= Remove["Global`*"]
```

Heat transfer in one dimension

Set up and solve the equations for different times

```
In[2]:= pde = D[T[y, u], u] == D[D[T[y, u], y], y];
        bcs = {T[0, u] == 0, T[1, u] == 100, T[y, 0] == T0};
        T0a = 50;
        T0b = 150;

In[5]:= sola = NDSolve[{pde, bcs /. T0 → T0a}, T[y, u], {y, 0, 1}, {u, 0, 10}];
        solb = NDSolve[{pde, bcs /. T0 → T0b}, T[y, u], {y, 0, 1}, {u, 0, 10}];
```

Separate plots for the two different initial temperatures

```
In[7]:= tempS = T[y, u] /. sola /. u → 0.1;
        tempM = T[y, u] /. sola /. u → 1;
        tempL = T[y, u] /. sola /. u → 10;
        Plot[{tempS, tempM, tempL}, {y, 0, 1},
         PlotStyle → {Solid, Dotted, Dashed}, PlotRange → {0, 150}]

In[11]:= tempS = T[y, u] /. solb /. u → 0.1;
         tempM = T[y, u] /. solb /. u → 1;
         tempL = T[y, u] /. solb /. u → 10;
         Plot[{tempS, tempM, tempL}, {y, 0, 1},
          PlotStyle → {Solid, Dotted, Dashed}, PlotRange → {0, 150}]
```

The natural length scale is ℓ, so define $y \equiv x/\ell$. This gives $\tau \equiv \ell^2/\kappa$ for the natural time scale, so define $u \equiv t/\tau$. In other words $u \ll 1$ ($u \gg 1$) corresponds to short (long) times. We therefore want to solve

$$\frac{\partial T}{\partial u} = \frac{\partial^2 T}{\partial y^2} \tag{5.7}$$

for $T(y, u)$ with $T(0, u) = 0$, $T(1, u) = 100$, and $T(y, 0) = T_0$.

The problem is very straightforward to set up in MATHEMATICA. Notebook 5.2 shows one approach. I define the differential equation and also the boundary conditions in terms of a parameter T_0. I then define two values for T_0, called a and b, and then use "NDSolve" to create a solution for each of a and b. (You may get a warning that the boundary conditions are inconsistent, caused by the discontinuities at the ends, but you can ignore that.)

Three curves (corresponding to "early", "middle", and "late" times) are generated for each plot. (I didn't bother to label the curves. You already know how to do that using the "PlotLegends" option, but for now, you can guess

which is which from context.) The two plots are shown below:

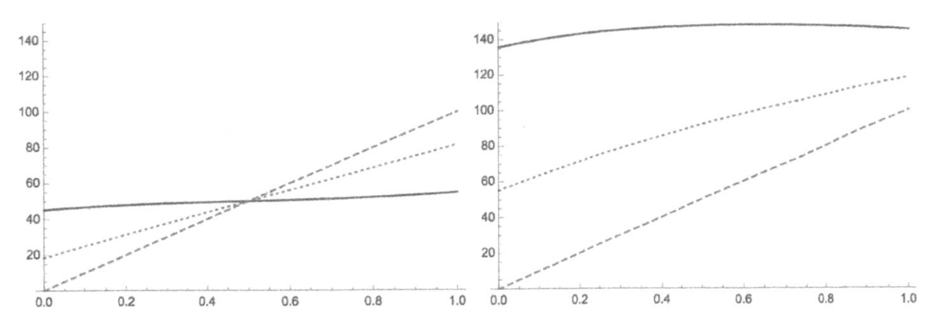

In each case, the "late" solution ($u = 10$) shows the same thing, namely a straight line from $T = 0$ at one end, to $T = 100$ at the other end. Also in each case, the "early" solution shows something close to a constant at $T = 50$ for a and at $T = 150$ for b. The "middle" line is a reasonable intermediate curve.

5.5 CHAPTER SUMMARY

- Try to rewrite differential equations in terms of the physical scales in the problem, so that functions and variables are dimensionless, before attempting numerical solutions to differential equations.
- Partial differential equations that are solvable analytically may require numerical solutions depending on the boundary conditions.
- Three-dimensional plots, using the functions "ContourPlot" or "DensityPlot", may be useful for displaying the solution of a partial differential equation.

EXERCISES

5.1 *A spring can deform if stretched long enough, leading to a restoring force*

$$F(x) = -kx\left[1 - \beta\left(\frac{x}{A}\right)^2\right]$$

where A is the maximum deformation and $|\beta|$ is small. If β is positive, the spring gets weaker the more it is extended, and the opposite is true if β is negative. Plot the motion for $x(0) = A$ and $\dot{x}(0) = 0$ over several periods for suitable values $\beta > 0$ and $\beta < 0$ and compare to $\beta = 0$. Choose values of A, k, and m which simplify the interpretation.

5.2 *A plane pendulum of length ℓ swings through an angle ϕ, where $\phi = 0$ is the vertical, equilibrium position. Solve for and plot the function $\phi(t)$, for t in units of $\tau = 2\pi(\ell/g)^{1/2}$, for a pendulum released from rest at $\phi(0) = \phi_0$, for a set of values for ϕ_0 ranging from something small to something close to $\phi_0 = \pi$. Compare the periods to the result from Example 3.2.*

5.3 *Consider an object with mass m that moves vertically close to the Earth's surface. It is subject to a drag force with magnitude βv^2 for a (positive) constant β where v is the object's velocity.*

a. *Let the object start from rest and fall from a height h. These conditions allow you to write the drag force in its simplest fashion. Write down and solve the differential equation analytically. Using $g = 9.8$ m/sec, $m = 0.25$ kg, $\beta = 0.01$ kg/m, and $h = 10$ m, find the time the object hits the ground and plot the height as a function of time up to that point.*

b. *Now write the drag force in a way that allows the object to move either upward or downward. This time, you can't solve the problem analytically, so solve it numerically, with the same values, and confirm that you get the same answers.*

c. *Now use initial conditions where the object starts at the ground and is given an initial upward velocity $v_0 = 30$ m/sec. Find again when it hits the ground, and plot the height as a function of time. Compare to the plot where there is no drag.*

5.4 *Transverse vibrations of a stretched string follow from the wave equation:*

$$\frac{1}{v^2}\frac{\partial^2 y}{\partial t^2} = \frac{\partial^2 y}{\partial x^2}$$

*where $y(x,t)$ is the shape of the string for a longitudinal position x at time t. Consider a string that has the shape $y(x,0) = (L/2 - x)^2(L/2 + x)^2$ at $t = 0$, where the string extends from $x = -L/2$ to $x = L/2$ and is fixed at the endpoints, that is $y(\pm L/2, t) = 0$. Also assume the string is not moving at $t = 0$. Find $y(x,t)$ and plot for several times, assuming some values for L and v. You can specify the initial condition that the string does not move with something like "*Derivative[0, 1][y][x, 0] == 0*". (This shape has no sharp changes in shape, and this is useful for achieving a numerical solution to the differential equation with default options.)*

5.5 *Set up the differential equations for motion in the xy-plane under an attractive central force $f = k/r^\beta$, and initial conditions $x(0) = a$, $y(0) = 0$, $\dot{x}(0) = 0$, and $\dot{y}(0) = v_0$. Pick some values for m, k, and a. First solve the equations for $\beta = 2$, and $v_0 = v_{\text{Circ}}$, the speed you calculate for a circular orbit. Use "*ParametricPlot*" to convince yourself the orbit is circular. Execute the solution for at least two circular orbit periods to convince yourself that the orbit is in fact closed. Next, use v_0 below and above the circular values, and convince yourself that you see closed elliptical orbits. Then choose $\beta = 3$ and again check for closed circular orbits when $v_0 = v_{\text{Circ}}$. Investigate what happens when you move away from the initial circular velocity.*

CHAPTER 6

Vectors and Matrices

CONTENTS

You probably encountered vectors in your very first physics course, as soon as you discussed motion in more than one dimension. Position in three dimensional space would be represented by something like $\mathbf{r} = (x, y, z)$ and velocity with $\mathbf{v} = (v_x, v_y, v_z)$. Other common vectors include force $\mathbf{F}$, electric field $\mathbf{E}$, and magnetic field $\mathbf{B}$.

This chapter shows you how to define vectors, and how to work with operations on them and between them. For example, the work dW done by a force $\mathbf{F}$ through a displacement $d\mathbf{r}$ is $dW = \mathbf{F} \cdot d\mathbf{r}$, and the force on a charge q in an electromagnetic field is $\mathbf{F} = q\mathbf{E} + q\mathbf{v} \times \mathbf{B}$. You will see that it is simple to use MATHEMATICA to apply these definitions to physical problems.

A matrix is a generalization of a vector. Like vectors, matrices are also useful for representing physical quantities. This chapter will review some of these physics concepts, and show you how to use MATHEMATICA to represent them and work with them. One particularly useful application is the product of a matrix and a vector, which produces another vector.

The mathematical concepts of eigenvalues, eigenvectors, and matrix diagonalization turn out to be especially relevant in physics. MATHEMATICA provides a suite of functions that make these manipulations straightforward, and this chapter will also go through these.

We are going to be working a lot with lists in this chapter. You should review Section 2.2 where lists were first discussed. We will also use the opportunity in this chapter to take a slight detour and discuss logical expressions and how to use them.

6.1 VECTORS AND MATRICES AS LISTS

Vectors and matrices are not entities within MATHEMATICA. Rather, we can treat list elements as if they were vectors or matrices. For example

```
v = {x, y, z}
```

just defines a variable "v" as a list of three elements called "x", "y", and "z". As far as MATHEMATICA is concerned, this is just another list. For example, "2v" returns "{2 x, 2 y, 2 z}", "v^2" returns "{x^2, y^2, z^2}", and "1/v" returns "{1/x, 1/y, 1/z}".

On the other hand, we can treat "v" as a three component vector. In this case, "2v" makes sense, but the results of "v^2" and "1/v" are nonsense. The meaning is all in the context of your usage.

MATHEMATICA gives you a convenient way to present vectors (and matrices) in a more familiar way. The output of "MatrixForm[v]" is

$$\begin{pmatrix} x \\ y \\ z \end{pmatrix}$$

That is, a column vector, where I again use the "TeXForm" command to create the output as printed here.

Matrices[1] are just nested lists. For example,

```
m = {{a, b, c}, {d, e, f}, {g, h, i}};
MatrixForm[m]
```

produces the output

$$\begin{pmatrix} a & b & c \\ d & e & f \\ g & h & i \end{pmatrix}$$

Of course, we can use "Part", or its shorthands, to extract specific elements of vectors or matrices. Sometimes you may want to perform operations on the elements of vector or matrix, that is, an operation on the elements of a list. Many such operations are described simply as list operations. For example, to make a sum of elements, you can execute

```
vec = {a, b, c, d};
len = Length[vec];
Sum[Part[vec, i], {i, 1, len}]
```

which simply returns "a + b + c + d". A simpler, but less flexible, way to get the same result is

```
vec = {a, b, c, d};
Total[vec]
```

We will see uses of both in one of the Physics Examples in Section 6.6.

[1] I will only use square matrices in this book.

6.2 LOGICAL EXPRESSIONS AND OPERATIONS

Before getting into the details of MATHEMATICA operations for vectors and matrices, we again take a detour to become familiar with the larger picture. This time, we briefly cover working with logical expressions.

We have already encountered logical expressions, in "Solve" (Chapter 2) and "DSolve" (Chapter 4), for example, but we did not begin to explore their full power. Consider

```
eq = x^2 == a^2;
sol = Solve[eq, x]
```

which returns

```
{{x -> -a}, {x -> a}}
```

The assignment "eq = x^2 == a^2" defines a variable "eq" as a logical statement, namely an equation. If we then execute

```
eq /. sol
```

we are returned

```
{True, True}
```

In other words, if we replace the variable x in the equation $x^2 = a^2$ with either the values $-a$ or $+a$, the equation is *True*. That is, the solution is correct. This is a very elegant way of demonstrating that you have the right solution to an equation.

In addition to *equals*, i.e. "==", the other relational operators are also available. As you expect, ">", "<", ">=", and "<=" mean *greater than*, *less than*, *greater than or equal to*, and *less than or equal to*. There is also "!=" which means *not equal to*.

MATHEMATICA allows operations among logical entities. The operators "&&", "||", and "!" are the logical operators *and*, *or*, and *not*. For example, executing the cell

```
True && False
True || False
! False
```

returns the values "False", "True", and "True".

That's enough for now. We'll make more use of these in the remaining pages of this book.

6.3 VECTOR OPERATIONS

Let's work with a pair of vectors:

```
v1 = {x1, y1, z1};
v2 = {x2, y2, z2};
```

Of course, any operation you would do for a list that makes sense for a vector, is a valid vector operation. For example, you can take the linear combination of any two vectors the obvious way, that is

```
a v1 - bv2
```

returns "{-bv2 + a x1, -bv2 + a y1, -bv2 + a z1}".

The inner product, aka dot product or scalar product, of two vectors is calculated using "Dot", and the cross product with "Cross". Specifically

```
Dot[v1, v2]
```

returns "x1 x2 + y1 y2 + z1 z2", and

```
Cross[v1, v2]
```

returns "-y2 z1 + y1 z2, x2 z1 - x1 z2, -x2 y1 + x1 y2". Both "Dot" and "Cross" have popular shorthand forms, using "." and "×". (The latter is typed in using "<Esc>cross<Esc>".) Try practicing your logical constructs by executing "Dot[v1, v2] == v1.v2" and "Cross[v1, v2] == v1 × v2", both of which should return with "True".

There is a function "Norm" which returns the length, or norm, of a vector:

```
Norm[v1]
```

This returns "$\sqrt{\text{Abs[x1]}^2 + \text{Abs[y1]}^2 + \text{Abs[z1]}^2}$" because MATHEMATICA realizes that the elements of the vector could be complex numbers. As we saw on Page 20, however, you can "Simplify" the result by assuming that the elements are all real numbers. That is

```
Simplify[Norm[v1], Assumptions -> {{x1, y1, z1} \[Element] Reals}]
```

returns "$\sqrt{\text{x1}^2 + \text{y1}^2 + \text{z1}^2}$".

There are some other vector operations, and I invite you to explore them on your own.

6.4 MATRIX OPERATIONS

Operations on matrices parallel operations on vectors where they make sense. That is, for two matrices "m1" and "m2", and two vectors "v1" and "v2",

a m1	means	multiply the matrix by a value
a m1+ b m2	means	form the linear combination
Dot[m1,m2]	means	matrix multiplication
m1.m2	means	the same thing as Dot[m1,m2]
m1.v1	means	multiply a matrix by a (column) vector
v1.m1	means	multiply a (row) vector by a matrix
v1.m1.v2	means	create a scalar

Note that there is no need to distinguish between "row" and "column" vectors.

There are a host of matrix operation commands that do exactly what you'd expect them to do. These include Transpose, ConjugateTranspose (aka *Hermitian adjoint*), Tr, (Trace), Det, (Determinant), and Inverse, as well as several others perhaps not as relevant for physicists.

MATHEMATICA also gives you tools for building matrices (and vectors). (That is, it gives you tools for building lists, which can be interpreted as matrices or vectors.) One of these tools is "Table[f,{i,n},{j,n}]" which builds an $n \times n$ square matrix with elements of the expression "f" evaluated over the given ranges of "i" and "j". For example

```
MatrixForm[Table[i + 2 j, {i, 3}, {j, 3}]]
```

produces the output

$$\begin{pmatrix} 3 & 5 & 7 \\ 4 & 6 & 8 \\ 5 & 7 & 9 \end{pmatrix}$$

Evidently, $i = 1, 2, 3$ labels the rows, and $j = 1, 2, 3$ labels the columns.

There is a built-in definition for the identity matrix of arbitrary dimension. Specifically "IdentityMatrix[n]" defines a list that can be interpreted as a $n \times n$ square matrix with ones along the diagonal and zeroes elsewhere.

For a final example, we can put together a few different things we've learned into the cell

```
m = {{a, b, c}, {d, e, f}, {g, h, i}};
mInv = Inverse[m];
Simplify[Dot[m, mInv]] == IdentityMatrix[3]
```

Executing this cell simply returns "True".

6.5 EIGENVALUE PROBLEMS

Eigenvalue problems probably represent the most important way that matrices are used in Physics. They are particularly important in classical mechanics and quantum mechanics.

Let's review briefly the mathematics. Think of an $n \times n$ matrix M which "operates" on a vector v, by multiplying the matrix times the vector. If the result is just the same vector v, perhaps multiplied by some value λ, then we say that λ is an eigenvalue corresponding to the eigenvector v. That is

$$Mv = \lambda v$$

There will be n eigenvectors, each with a corresponding eigenvalue. In mechanics, this mathematics shows up when analyzing systems of coupled oscillators ("eigenfrequencies") and the rotation of rigid bodies ("principal moments of inertia"). One of the fundamental postulates of quantum mechanics is based on the eigenvalues of "operators" that correspond to observable quantities.

Finding the eigenvalues and eigenvectors of a particular matrix is simple in MATHEMATICA. The command "Eigenvalues[m]" returns the eigenvalues of the matrix "m", and "Eigenvectors[m]" returns the corresponding eigenvectors. There is also "Eigensystem[m]" which returns a combined list of the eigenvectors and eigenvalues.

Let's do a simple example to see how this works. Execute the cell

```
m = {{0, a}, {a, 0}};
eivals = Eigenvalues[m]
eivecs = Eigenvectors[m]
eiboth = Eigensystem[m]
```

and the commands will return

```
{-a, a}
{{-1, 1}, {1, 1}}
{{-a, a}, {{-1, 1}, {1, 1}}}
```

You can test these answers by showing that "m.{1, 1}" returns "{a, a}", and "m.{-1, 1}" returns "{a, -a}". That is

$$\begin{pmatrix} 0 & a \\ a & 0 \end{pmatrix}\begin{pmatrix} 1 \\ 1 \end{pmatrix} = a\begin{pmatrix} 1 \\ 1 \end{pmatrix} \quad \text{and} \quad \begin{pmatrix} 0 & a \\ a & 0 \end{pmatrix}\begin{pmatrix} -1 \\ 1 \end{pmatrix} = -a\begin{pmatrix} -1 \\ 1 \end{pmatrix}$$

It frequently happens in Physics problems, that you instead encounter the Generalized Eigenvalue Problem, which involves a second matrix Q:

$$Mv = \lambda Q v \tag{6.1}$$

In this case, use the syntax "Eigenvectors[{m,q}]" to include the second matrix "q". (Note the brackets in the argument!) Beware, however, that the complete symbolic solution in this case can be quite complicated, and you may want to resort to numerical computation.

Before moving on to the physics examples, let's take a moment to talk about "diagonalization" and how we can carry it out in MATHEMATICA. We can illustrate this with a simple example using a 2×2 symmetric matrix, set up to make the numbers come out neatly. Executing

```
m = {{2, Sqrt[15]}, {Sqrt[15], 4}};
\[Lambda] = Eigenvalues[m]
v = Eigenvectors[m]
```

returns the eigenvalue list

$$\{7, -1\}$$

and the list of eigenvectors

$$\left\{\left\{\sqrt{\frac{3}{5}}, 1\right\}, \left\{-\sqrt{\frac{5}{3}}, 1\right\}\right\}$$

The eigenvectors of a symmetric matrix must be orthogonal to each other, and it is easy enough to check that "v[[1]].v[[2]]" returns zero. Note, however, that the function "Eigenvectors" does not, necessarily, return eigenvectors that are normalized to any particular value.

The format returned by "Eigenvectors" is a list of lists, that is, a list of the eigenvectors. The larger list, therefore, can be interpreted as a matrix whose rows are the eigenvectors. For example, continuing from above, executing

```
vT = Transpose[v];
v.vT
```

returns

$$\left\{\left\{\frac{8}{5}, 0\right\}, \left\{0, \frac{8}{3}\right\}\right\}$$

which is exactly what you'd expect. This operation forms the inner product between eigenvectors that are the rows of "v" and the columns of "vT". The diagonal elements are just the squares of the norms of the two eigenvectors.

In order to use "v" as a transformation matrix that diagonalizes "m", the eigenvectors need to be normalized. A brute force way to do that is

```
u = Table[Normalize[v[[i]]], {i, Length[v]}];
```

where "Normalize" returns a normalized version of the vector in its argument. A cleaner way to do this uses the "loop" feature of "Table", that is

```
u = Table[Normalize[ev], {ev, v}]
```

However, the most succinct way of normalizing the eigenvectors uses the "Map" function, or its shorthand. These are

```
u = Map[Normalize, v]
```

or

```
u = Normalize /@ v
```

which "map" the function "Normalize" onto the vector "v". We can now diagonalize the original matrix by executing

```
u = Normalize /@ v;
MatrixForm[u.uT]
MatrixForm[uT.u]
MatrixForm[Simplify[u.m.uT]]
```

which returns the three matrices

$$\begin{pmatrix} 1 & 0 \\ 0 & 1 \end{pmatrix} \quad , \quad \begin{pmatrix} 1 & 0 \\ 0 & 1 \end{pmatrix} \quad \text{and} \quad \begin{pmatrix} 7 & 0 \\ 0 & -1 \end{pmatrix}$$

where the first two matrices just prove that "u" is indeed unitary, and the third matrix is the diagonalized form of "m".

Of course, all this works symbolically as well. Be aware, however, that in many cases, the forms of the eigenvalues and eigenvectors are complicated and do not reduce easily, if at all, with functions like "Simplify".

6.6 PHYSICS EXAMPLES

Example 6.1 *Four masses m lie in the xy plane at the corners of a rectangle as shown in Figure 6.1, with one of the masses located at the origin. Find the location of the center of mass, and the principal axes and moments of inertia, about axes located at the center of mass.*

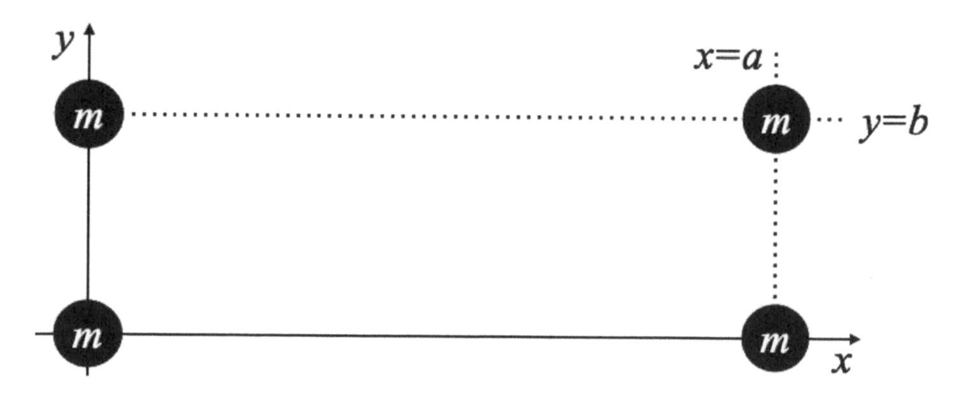

FIGURE 6.1 Diagram to go with Example 6.1

This is a bit of an advanced topic, so let's take a moment to review the physics and relevant definitions. Given a rigid body made of a collection of N masses m_α located at position vectors $\mathbf{r}_\alpha$ that are fixed relative to each other, the vector that locates the *center of mass* is given by

$$\mathbf{r}_{\rm CM} = \frac{\sum_{\alpha=1}^{N} m_\alpha \mathbf{r}_\alpha}{\sum_{\alpha=1}^{N} m_\alpha} \tag{6.2}$$

That is, it is the average location of all the positions weighted by the individual masses. You might refer to this as the "first moment" of the mass.

The components of the inertia tensor are defined by

$$I_{ij} = \delta_{ij} \left[\sum_{\alpha=1}^{N} \left(\sum_{k=1}^{3} m_\alpha r_{\alpha,k}^2 \right) \right] - \sum_{\alpha=1}^{N} (m_\alpha r_{\alpha,i} r_{\alpha,j}) \tag{6.3}$$

which has the appearance of a "second moment" of the mass. The factor

$$\sum_{k=1}^{3} r_{\alpha,k}^2 = \mathbf{r}_\alpha \cdot \mathbf{r}_\alpha = r_\alpha^2$$

in the first term is just the length-squared of the vector $\mathbf{r}_\alpha$. The second term of Equation 6.3 contains elements of something called an outer product $\mathbf{r}_\alpha \mathbf{r}_\alpha$. That is, we can write the inertia tensor as

$$I = \sum_{\alpha=1}^{N} (m_\alpha r_\alpha^2 \, 1 - m_\alpha \mathbf{r}_\alpha \mathbf{r}_\alpha) \tag{6.4}$$

where 1 is the identity tensor.

You might first encounter the *inertia* tensor when connecting the angular momentum vector $\mathbf{L}$ and the angular velocity vector $\boldsymbol{\omega}$. You might be under the impression from freshman physics that these two vectors are parallel, but they are not. They are related by an inertia tensor I, which is a 3×3 matrix when written in terms of components. That is, where $i = 1, 2, 3$ means x, y, z,

$$L_i = \sum_{j=1}^{3} I_{ij} \omega_j$$

It is possible to rotate your coordinate axes to one in which I is diagonal, in other words[2] $I_{ij} = I_i \delta_{ij}$ and $L_i = I_i \omega_i$. That is, any component of the angular momentum is the corresponding diagonal element times that component of the angular velocity. Physically, $\mathbf{L}$ is now parallel to $\boldsymbol{\omega}$. Mathematically, the product $I \cdot \boldsymbol{\omega}$ is replaced by one of three numbers times the corresponding element of $\boldsymbol{\omega}$. That is, we have found the eigenvalues of the matrix corresponding to I, and the eigenvectors are the corresponding directions, called the *principal axes.*

[2]The Kronecker delta δ_{ij} is unity for $i = j$ but zero otherwise.

Now let's move on to the problem solution. In fact, the solution is obvious because as stated, there is plenty of symmetry. The center of mass is at $(x, y) = (a/2, b/2)$, the principal axes are just the x and y axes, and the corresponding moments of inertia are $4 \times m(b/2)^2 = mb^2$ and $4 \times m(a/2)^2 = ma^2$. However, we can set up the problem for four arbitrary masses or four arbitrary positions, but solve it for this special case and check the answer.

Notebook 6.1 shows my solution. We'll go through it slowly, step by step, and I encourage you to try it yourself, removing the semicolons and inspecting the output. You should also play with different arrangements of masses and convince yourself that you get the correct result.

The calculation only works in two dimensions, but you can add a third dimension if you like.

First we calculate the center of mass location. The four position vectors are stored in "rVecs" as a nested list. (You could generalize this to four arbitrary positions (x_α, y_α) and then replace these values with these to check the answer, and a different set for a different answer.) The four masses m_α are also stored in a list, although here the four values are the same. In preparation for calculating Equation 6.2, we store in "mr" the masses and the vectors, and in "mTot" the sum of the masses.

The calculation of "mTot" and the CM coordinates uses the function "Total", which sums over the (inner) list elements. That is, the total mass is just the total of the four masses, and the CM coordinates are the total of the masses times their coordinate vectors, divided by the total mass. The result $(x_{\rm CM}, y_{\rm CM}) = (a/2, b/2)$ is correct.

Next we move on to the inertia tensor. We construct a vector "rCM" by subtracting off from "rVecs" the coordinates of the center of mass. To do this, we use "Table" to create a list of vectors of center of mass coordinates. For use in the next step, we store in the lists "mrCM" and "mrCM2" the products of the masses m_α times the center-of-mass coordinate vectors $\mathbf{r}_\alpha$, and a construct that is m_α times the square of every element of the list "rCM".

Now we are ready to calculate the two terms of the inertia tensor in Equation 6.4. The first term is just a number $\sum_\alpha m_\alpha r_\alpha^2$ times the identity matrix. This number is calculated and stored in "iTerm1" by first summing over the mass m_α times the squares of the coordinates x_α and y_α, and then summing these terms over α. (The summations are done using "Total".) This result is then multiplied by the 2×2 identity matrix to get the first term of the inertia tensor.

The second term "iTerm2" is calculated in a similar fashion, but we need to calculate the outer product $m_\alpha \mathbf{r}_\alpha \mathbf{r}_\alpha$ in Equation 6.4. To do this, we use the MATHEMATICA command "KroneckerProduct". (There is a command "Outer" that is considerably more general.) This takes the two-dimensional vector for each mass α and forms the outer product. I then use "Sum" to explicitly sum over α.

The command "Eigensystem" indeed returns the right answer. However, you should confirm that, in fact for this simple example, the inertia tensor is

NOTEBOOK 6.1 Solution to Example 6.1.

```
In[1]:= Remove["Global`*"]
```

Four masses in a plane

Calculate the center of mass

```
In[2]:= rVecs = {{0, 0}, {a, 0}, {0, b}, {a, b}};
        mVals = {m, m, m, m};
        mr = mVals rVecs;
        mTot = Total[mVals];
        {xCM, yCM} = Total[mr] / mTot
```

Out[6]= $\left\{\frac{a}{2}, \frac{b}{2}\right\}$

Form the inertia tensor relative to CM

```
In[7]:= rCM = rVecs - Table[{xCM, yCM}, 4];
        mrCM = mVals rCM;
        mrCM2 = mVals rCM^2;
        iTerm1 = Total[Total[mrCM2]];
        iTerm2 = Sum[
            KroneckerProduct[
             Part[mrCM, alpha], Part[rCM, alpha]
            ],
            {alpha, 1, 4}];
        inertia = iTerm1 IdentityMatrix[2] - iTerm2;
```

Get the eigenvalues and eigenvectors

```
In[13]:= Eigensystem[inertia]
```

Out[13]= $\left\{\left\{a^2 m, b^2 m\right\}, \{\{0, 1\}, \{1, 0\}\}\right\}$

already diagonal, but here is where you should try different combinations and check the result.

Example 6.2 *Reconsider Exercise 4.3 by assuming the form of a solution which contains a single frequency ω, and use this to find possible values of ω.*

We are asked to look for solutions to Equations 4.3 that have a single frequency ω, so write the solution in the form

$$x_1(t) = a_1 e^{i\omega t} \qquad \text{and} \qquad x_2(t) = a_2 e^{i\omega t} \tag{6.5}$$

This gives

$$\begin{aligned} -m\omega^2 a_1 &= -ka_1 + k(a_2 - a_1) \\ -m\omega^2 a_2 &= -ka_2 - k(a_2 - a_1) \end{aligned}$$

If we write $\omega_0^2 \equiv k/m$, then we write these equations in matrix form as

$$\begin{pmatrix} 2 & -1 \\ -1 & 2 \end{pmatrix} \begin{pmatrix} a_1 \\ a_2 \end{pmatrix} = \frac{\omega^2}{\omega_0^2} \begin{pmatrix} a_1 \\ a_2 \end{pmatrix}$$

This is just an eigenvalue equation where $\lambda \equiv \omega^2/\omega_0^2$ is the eigenvalue. Using "Eigensystem[{{2, -1}, {-1, 2}}]", which returns "{{3, 1}, {{-1, 1}, {1, 1}}", tells us that the eigenfrequencies are $\omega = \sqrt{3}\omega_0$ and $\omega = \omega_0$. The corresponding modes are when the two masses opposite against each other ($-a_1 = a_2$) and parallel to each other ($a_1 = a_2$). This is just the answer we found in Exercise 4.3.

Now let's consider something a little more challenging.

Example 6.3 *Two masses m_1 and m_2 are connected by three springs with stiffness k_1, k_2, and k_3 to each other and to fixed walls on either side of a frictionless, horizontal surface, as shown in Figure 6.2. Find the eigenfrequencies and eigenmodes of oscillation. Analyze the special case $m_1 = m_2$, $k_1 = k_3$, and $k_2 = rk_1$. Show that your answer agrees with the special case in Exercise 6.2 when $r = 1$. Discuss the solutions for $r \ll 1$ and $r \gg 1$.*

You should be able to follow Example 4.3 to set up the coupled equations of motion, substitute the Ansatz 6.5, and arrive at the equation

$$Ka = \omega^2 Ma \tag{6.6}$$

where $K = \begin{pmatrix} k_1 + k_2 & -k_2 \\ -k_2 & k_2 + k_3 \end{pmatrix}$, $M = \begin{pmatrix} m_1 & 0 \\ 0 & m_2 \end{pmatrix}$, and $a = \begin{pmatrix} a_1 \\ a_2 \end{pmatrix}$.

This is a generalized eigenvalue equation of the form (6.1), and a solution is shown in Notebook 6.2.

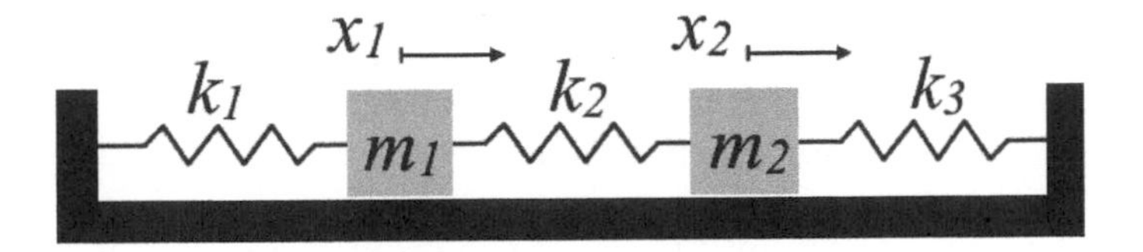

FIGURE 6.2 Diagram to go with Example 6.3

NOTEBOOK 6.2 Solution to Example 6.3.

```
In[1]:= Remove["Global`*"]
```

General 2m 3k problem

Define the matrices

```
In[2]:= kM = {{k1 + k2, -k2}, {-k2, k2 + k3}};
        mM = {{m1, 0}, {0, m2}};
```

Find the eigenvalues and eigenvectors

```
In[4]:= {vals, vecs} = Eigensystem[{kM, mM}];
```

Check the answers

```
In[5]:= rep = {m1 → m, m2 → m, k1 → k, k2 → r k, k3 → k};
        assmp = {k > 0, m > 0, r > 0}
        Simplify[vals /. rep, Assumptions → assmp]
        Simplify[vecs /. rep, Assumptions → assmp]
Out[6]= {k > 0, m > 0, r > 0}
```

Out[7]= $\{\frac{k}{m}, \frac{k + 2 k r}{m}\}$

```
Out[8]= {{1, 1}, {-1, 1}}
```

Except for the use of "Eigensystem" for the generalized eigenvalue problem, you have seen everything in this simple notebook. Notice that in this case, a list "{vals,vecs}" is used to accept the return from "Eigensystem". You should take the time to remove the semicolon and see the result, although it is rather complicated, and see if you can make sense of it.

To check the answers, we set $m_1 = m_2 = m$, $k_1 = k_2 = k$, and $k_2 = rk$, which gives us the opportunity to check the result if $r = 1$ or if the middle spring is either very weak ($r \ll 1$) or very strong ($r \gg 1$).

Firstly, regardless of the value of r, the eigenmodes are the same as in Exercise 6.2. That is, they correspond to the masses moving together, i.e. "vecs" equal to "{1,1}", or moving in the opposite directions with the same amplitude, i.e. "vecs" equal to "{-1,1}". This is just a consequence of the symmetry, with equal masses and spring constants on the outside.

For $r = 1$, the two eigenfrequencies are $\sqrt{k/m}$ and $\sqrt{3k/m}$, just as we found in Exercise 6.2. In fact, one of the eigenfrequencies is $\sqrt{k/m}$ for any value of r, corresponding to the eigenmode when the two masses oscillate together. The middle spring is irrelevant since it never changes length. The problem is equivalent to a mass $2m$ connected to a spring with stiffness $2k$.

For a very weak spring k_2, $r \to 0$ and the two eigenfrequencies are the same. We essentially have two independent oscillators, although there can be interesting effects if r is small but finite. This will cause energy to be transferred from one oscillator to the other and back again. See Exercise 6.5.

If the middle spring is very stiff, $r \to \infty$, the system is essentially two masses (with reduced mass $\mu = m/2$) connected by a spring with stiffness $rk = k_2$, so the frequency is $\sqrt{k_2/\mu} = \sqrt{2rk/m}$ which again agrees with our result.

We presented Physics Examples 6.2 and 6.3 by looking for a solution with a "single frequency", but what motion is represented by that single frequency? Chapter Exercise 6.5 makes it clear that the motions of the individual masses are a combination of frequencies. It is the linear combination of their motions, given by the eigenvectors, that moves at a single frequency. These linear combinations are called *Normal coordinates* and the coefficients of the linear combination are the mass-weighted eigenvector components.

The theory of small, coupled oscillations can be found in any intermediate or advanced textbook on classical mechanics, and we won't cover it here. See, however, Exercise A.2 in Appendix A for one example. You might also modify Exercise 6.3 and find the right linear combination for different masses.

6.7 CHAPTER SUMMARY

- In MATHEMATICA, vectors and matrices are just lists that are interpreted in special ways. The function "MatrixForm" can be used to output these lists in the familiar form.
- In addition to normal list operations, for example addition, there are

special operations that pertain to vectors. Examples of these include "Dot", "Cross", and "Norm".

- These special operations are naturally extended to matrices.
- A full suite of functions are provided for solving eigenvalue problems.
- This chapter also gives you more examples using logical expressions, and illustrates them using matrix manipulations.

EXERCISES

6.1 *A mass m hangs vertically from a thin, massless wire. The top of the wire is attached to one end of a horizontal strut. The other end of the strut abuts a vertical wall. A second thin, massless wire holds the strut horizontal, making an angle θ with the strut, as shown below:*

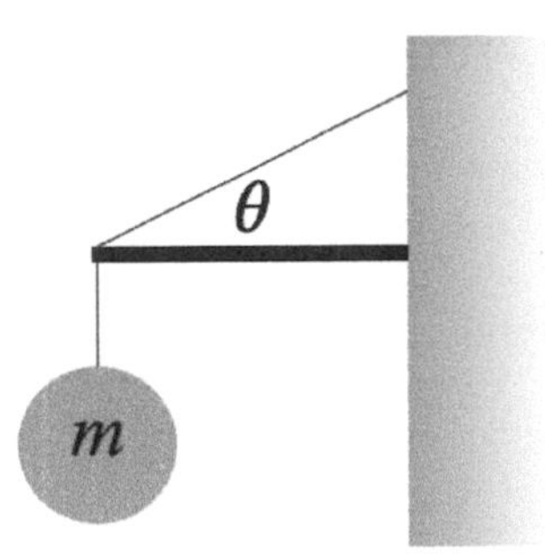

Find the compression force on the strut in terms of m, θ, and the acceleration g due to gravity. Make a plot of this force as a function of θ for $m = 1$ kg.

6.2 *Two equal masses m are fixed to the end of a straight, rigid, massless rod of length 2ℓ. The rod makes an angle θ with the z-axis and rotates with an angular velocity ω about a fixed point at the origin, as shown below.*

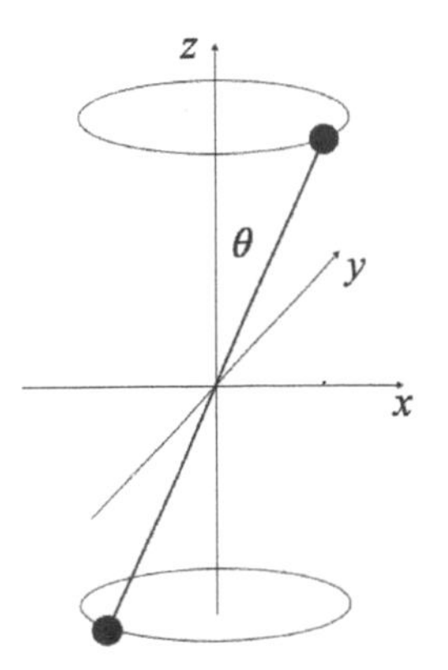

The two masses are located at $\mathbf{r}_1$ *and* $\mathbf{r}_2$*, and their velocities are* $\mathbf{v}_{1,2} = \dot{\mathbf{r}}_{1,2}$*. Show that* $\mathbf{v}_{1,2} = \boldsymbol{\omega} \times \mathbf{r}_{1,2}$ *where* $\boldsymbol{\omega} = \omega\hat{\mathbf{z}}$*. Then, construct the total angular momentum vector*

$$\mathbf{L} = \sum_{i=1,2} \mathbf{r}_i \times m\mathbf{v}_i$$

and use the inner product $\boldsymbol{\omega} \cdot \mathbf{L}$ *to find (in its simplest form) the angle* α *between the angular velocity and angular momentum.*

6.3 *A particle with mass* m *and charge* q *moves in a uniform magnetic field* $\mathbf{B} = B_0\hat{\mathbf{z}}$*, so it feels a force* $\mathbf{F} = q\mathbf{v} \times \mathbf{B}$*. It starts at time* $t = 0$ *at position* $(x, y, z) = (a, 0, 0)$ *with velocity vector* $(v_x, v_y, v_z) = (0, u, w)$*. Solve the simultaneous differential equations of motion to find* $\mathbf{r}(t) = (x(t), y(t), z(t))$*, and show that the particle's speed is constant (and equal to the initial value).*

6.4 *Find the center of mass, moments of inertia, and principal axes for three equal masses* m*. Two of the masses lie on the* x*-axis at* $x = \pm a/2$ *and the third is on the* y*-axis at* $y = b$*, forming an isosceles triangle.*

6.5 *Use Example 6.3 to find the motions of each of the two masses, for the case* $m_1 = m_2 \equiv m$ *and* $k_1 = k_3 \equiv k = 10k_2$*. Use initial conditions* $x_1(0) = A$ *and* $\dot{x}_1(0) = x_2(0) = \dot{x}_2(0) = 0$ *and make a plot of both functions versus time, demonstrating how the amplitude oscillates from* m_1 *to* m_2 *and back again.*

6.6 *Find the eigenfrequencies and eigenmodes of the mass and spring system shown below:*

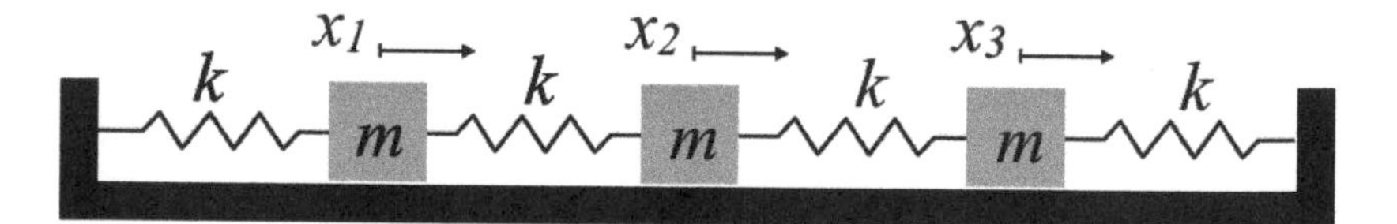

CHAPTER 7

Basic Data Analysis

CONTENTS

Data analysis is a fundamental activity in physics, as well as other sciences. This chapter shows you how to use MATHEMATICA to analyze numerical data.

For our purposes here, "data" will mean a set of numerical values that represent some measurement or measurements. For example, a list of numbers might be a series of measurements of the air temperature in a certain room, all taken under identical circumstances with a scatter in the values due to some experimental uncertainty.

We will also make use of "two column" data, where the first column is some control variable and the second a measured value, and this chapter covers this case as well. For example, we might want to analyze the temperature in a room recorded at different times of the day. There are more complex examples, but the manipulations and tools we use here will apply to them as well.

The MATHEMATICA functions which carry out these basic functions are covered in this chapter. For example, given a list of numbers, what is the maximum value? The minimum value? The average and the standard deviation of the values? How can you select a set of values that satisfy a specific criterion, perhaps selecting values from the second column based on criteria applied to the first?

Other tools discussed in this chapter include reading in data from a file, and making histograms (also known as "frequency plots") to graphically represent the data. More complex tools are discussed in later chapters.

The MATHEMATICA Documentation Center has a section on "Data Manipulation & Analysis", including a number of guides and tutorials such as "Handling Arrays of Data". This is a good place to go for additional details.

7.1 NUMBERS IN LISTS

Think of a data set as a list of numbers. In MATHEMATICA language,

```
data = {0, 9, 2, 8, 3, 6, -9, -7, -5, -1,
   -8.4, 9.3, -7.8, 6.0, -5.9, 4.1, -3.2, 1.9};
```

is an example of a data set. We will learn later how to read in data sets from files, get them from external sources, or generate them ourselves, but for now, the list above will be useful for demonstrating some basic concepts.

Basic information about the data set is simple to get. For example

```
Length[data]
Min[data]
Max[data]
```

return "18", "-9", and "9.3".

You can also easily find the sum of all the data values. Executing

```
len = Length[data];
Sum[Part[data, i], {i, 1, len}]
```

returns the value "2.", the "." explicitly indicating a floating point number. This is actually a good time to begin graduating to the "double bracket" shorthand for "Part", that is

```
len = Length[data];
Sum[data[[i]], {i, 1, len}]
```

returns the same as the above construction.

Various statistical analysis functions are readily available. For example

```
Mean[data]
StandardDeviation[data]
```

returns "0.111111" and "6.27384". You might confirm that

```
avg = Mean[data];
Sqrt[Sum[(data[[i]] - avg)^2, {i, 1, len}]/(len - 1)]
```

gives the same result as "StandardDeviation[data]".

It is straightforward to "Sort" a data set, that is

```
Sort[data]
```

returns

```
{-9, -8.4, -7.8, -7, -5.9, -5, -3.2, -1, 0,
        1.9, 2, 3, 4.1, 6, 6., 8, 9, 9.3}
```

where I have artificially split the line of output to fit on this page. A similar command "Union[data]" orders the data set and eliminates duplicate values, and can also be used to join two or more data sets into one ordered list.

A data set might consist of two or more correlated lists, for example

```
dataA = {-3.1, -0.5, 1.9, 4.6, 6.2};
dataB = {10.5, 13.2, 6.5, 15.3, 0.9};
```

There are a variety of built-in data analysis tools that allow you to combine lists and explore correlations, but you can also do a lot just by using simple list manipulation. For example

```
dataAB = {dataA, dataB}
```

creates a nested list of the two lists, and

```
dataABpaired = Transpose[dataAB]
```

turns this into a list of data pairs, one of the pair from "dataA" and the other from "dataB". Of course,

```
Transpose[dataABpaired] == dataAB
```

returns "True".

This is a good time to introduce you to an extended use of "Part", as well as the built-in object "All". So long as a list is deep enough, you can include additional indices as arguments to "Part" to extract specific elements. If you list an index as "All", then it includes all indices up to the length of the object. So, for example,

```
Part[dataABpaired, All, 1]
```

returns

```
{-3.1, -0.5, 1.9, 4.6, 6.2}
```

Alternatively,

```
Part[dataABpaired, 1, All]
```

returns

```
{-3.1, 10.5}
```

In shorthand, "dataABpaired[[All, 1]]" and "dataABpaired[[1, All]]", do the same thing, respectively.

There are different options for displaying data. "ListPlot[data]" produces

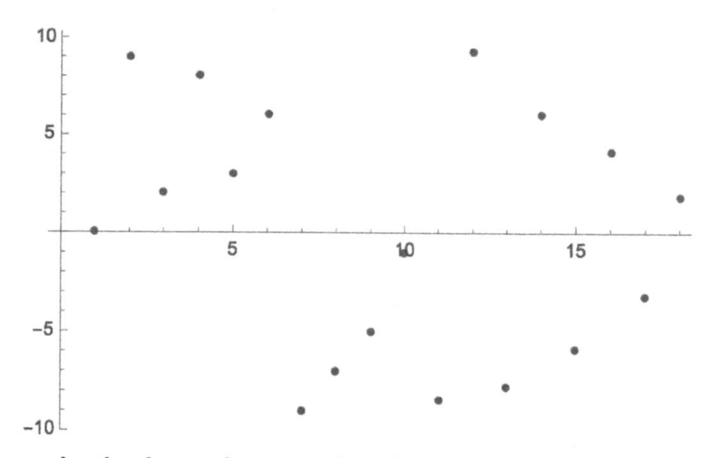

In other words, the data values are plotted as a function of their ordinal number in the list. Other, similar functions include "ListLinePlot", which connects the data values by lines, and "ListStepPlot" that draws steps at each value.

Correlated data lists can be plotted individually or as pairs of numbers. For example, the command

```
ListPlot[dataAB, PlotMarkers -> {"A", "B"}]
```

produces the plot

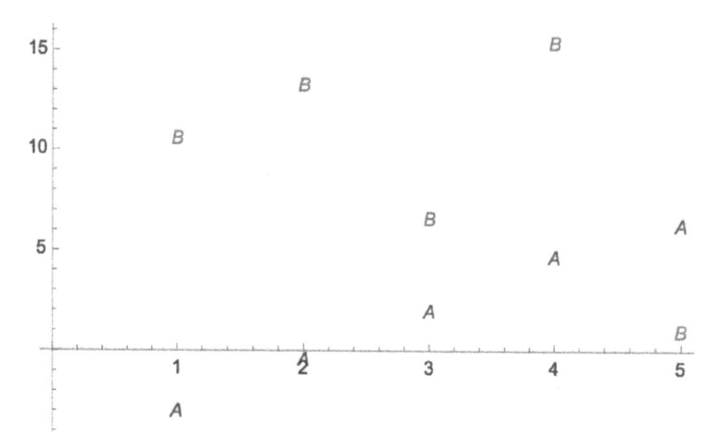

and the command

```
ListPlot[dataABpaired, PlotMarkers -> {"X", Large}]
```

creates

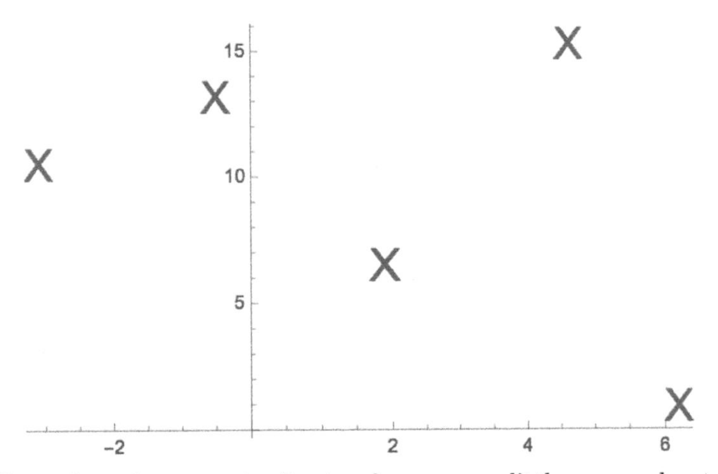

where I've taken the opportunity to show you a little more about how to embellish the output.

7.2 SELECTING DATA SEGMENTS

Physicists often want to select portions of a data set based on certain criteria.[1] This section shows you how to do this in MATHEMATICA.

The concept behind data selection in MATHEMATICA is simple, but the execution can be hard to follow at first. You should refer to the tutorial on "Testing Expressions", as well as our previous discussions of logical expressions (Secton 6.2) and pure functions (Page 25).

To illustrate the methods, we will make use of the simple data structures defined in Section 7.1. The MATHEMATICA function used to select data is "Select". Its arguments are the data set in question and a criterion (as a logical function) for making the selection. It returns the data subset of elements that meet the criterion. So, for example

```
Select[data, # > 0 &]
```

returns

```
{9, 2, 8, 3, 6, 9.3, 6., 4.1, 1.9}
```

That is, it returns only the elements of "data" that meet the criterion specified by the pure function "# > 0 &", in other words, greater than zero.

More complicated functions are of course possible. The command

```
Select[data, Abs[#] > 5 &]
```

returns

```
{9, 8, 6, -9, -7, -8.4, 9.3, -7.8, 6., -5.9}
```

[1]We often refer to this as "making cuts" on the data set.

and the command

```
Select[data, # > 0 && # < 6 &]
```

returns

```
{2, 3, 4.1, 1.9}
```

There are also some built-in logical functions that can test values, including "IntegerQ", "PrimeQ", and "Positive". So, for example

```
Select[data, Positive] == Select[data, # > 0 &]
```

returns "True".

Now suppose we have a data set made of nested lists of data pairs and we want to extract the second values of the pairs, using a selection based on the first values. For example, referring to the data set "dataABpaired" defined in Section 7.1, extract the B values for which the corresponding A value is greater than zero. The answer should be the three numbers 6.5, 15.3, and 0.9. How do we carry out this selection? Any of the four commands

```
Part[Select[dataABpaired, Part[#, 1] > 0 &], All, 2]
Part[Select[dataABpaired, #[[1]] > 0 &], All, 2]
Select[dataABpaired, Part[#, 1] > 0 &][[All, 2]]
Select[dataABpaired, #[[1]] > 0 &][[All, 2]]
```

give the correct result, making use of the shorthand for "Part" in none, one, or both of the two places it is needed. This is, in principle, clear from what we have covered so far in this chapter, but it is good for you to try all this out and convince yourself that you understand the syntax.

7.3 READING DATA FROM A FILE

Much more often than not, the data you want to analyze will be stored in a file somewhere, and you don't want to take the time (and risk the error) of re-entering that data by hand. Instead, you want to read the data directly from the file.

The MATHEMATICA function which does this is "Import". This function does everything it can to understand your data format, and return the appropriate object, but you can also specify the format if you like. We will only work with the simplest data file formats in this book, but you should review the MATHEMATICA documentation, particularly the section on "Data Manipulation", to learn more.

Even if you can get away without specifying the data format, you still need to tell MATHEMATICA where the data file is located. This you can do with the "SetDirectory" function. There are also built-in symbols that can help you identify the relevant directory. If you are executing a notebook, you can specify the directory in which it resides using the function "NotebookDirectory".

Let's illustrate this with an example. A file[2] containing a carriage-return

[2]The file "rangs.dat" was written with the code discussed in Section 10.3.

separated list of 1000 random numbers, distributed according to a Gaussian with mean 1.5 and standard deviation 2.5. The file sits in the directory "DataFiles" which is just above the directory where the notebook is executing. To read the file, execute the cell

```
SetDirectory[NotebookDirectory[]];
SetDirectory["./DataFiles"];
rg = Flatten[Import["rangs.dat"]];
```

The "Flatten" function is used to make "rg" into a list oriented as a single list of numbers. Then "ListPlot[rg]" produces the following:

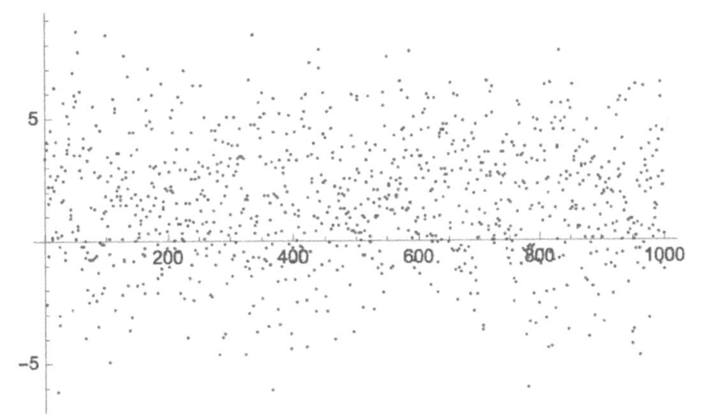

The commands "Mean[rg]" and "StandardDeviation[rg]" return reasonable values, based on the parameters used to generate the numbers in the file.

7.4 MAKING HISTOGRAMS

A histogram, or "frequency distribution", is useful way to display data. It plots the number of occurrences, that is the frequency, of data values, as a function of some distribution variable. In its simplest form, it shows the different measurements of the same variable, that have different values because of some kind of random or systematic variation.

MATHEMATICA provides "Histogram" to plot data as a histogram. It also provides "HistogramList" to return the numbers in the plot. Both functions take essentially the same arguments, except that one makes a plot and the other returns the numbers that make up the plot. The first argument is the data set to histogram, and the second argument specifies the bins. The bins can be specified by their number, their width, or explicitly as the lower and upper limit with the number of bins in between.

We can use the Gaussian distribution data set "rg" from Section 7.3 to illustrate how this works. The cell

```
bList = {-5, 8, 0.25};
```

```
Histogram[rg, bList]
vals = HistogramList[rg, bList];
ListPlot[vals[[2]]]
```

produces the two plots shown here:

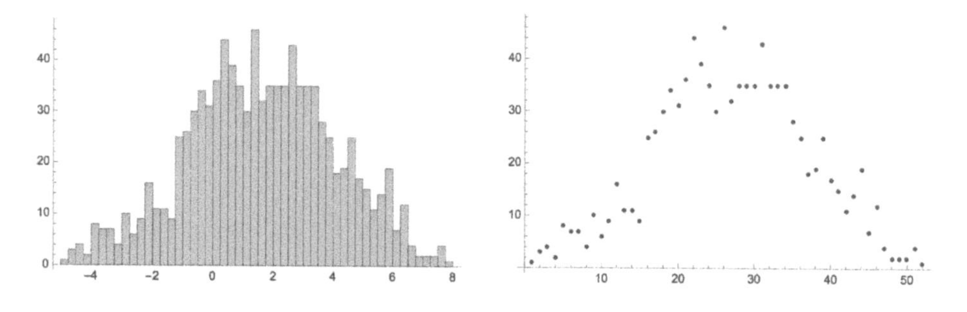

The object "bList" is the bin specification, binning the data between the values -5 and 8 in bins that are 0.25 units wide. The first plot comes from the "Histogram" command. It clearly shows something that peaks in the vicinity of $\mu = 1.5$ and has a standard deviation σ near 2.5.

The second plot is of the same data, where the frequencies are extracted into the object "vals" using "HistogramList", and plotted using "ListPlot". In fact, "vals" contains both the bin edges in the first part, and the actual frequencies in the second part. The plotted points are obviously of the same distribution as given by "Histogram", but the plot is as a function of the bin number, not the bin edges.

Many variations and plot options are possible, and you should explore them, but this is enough for now.

7.5 PHYSICS EXAMPLE

This one example encompasses all of the concepts in this chapter, and also gives you the chance to extend some of them with options and embellishments.

Example 7.1 *Does the gas mileage of a car vary between summer and winter? A two-column data file* mpg.dat *has data for a routinely maintained 1994 Honda Accord. The first column is the day since 1 July 2008, and the second is the gas mileage (in miles per gallon) on that date. Find the average gas mileage and its standard deviation, and plot the gas mileage versus time. Then separate the data sets into January/February and July/August. Histogram these two data sets, and find the average gas mileage and the standard deviation.*

My solution to this example is shown in Notebook 7.1. I have deleted the output to save space, but we will go through the notebook step by step, and I will include the plots and other results along the way.

The first step is to read the data file. As before, the file is in a subdirectory

NOTEBOOK 7.1 One possible solution to Example 7.1

```
In[1]:= Remove["Global`*"]
```

Data Analysis of Gas Mileage

Read in the data file and extract columns

```
In[2]:= SetDirectory[NotebookDirectory[]];
        SetDirectory["./DataFiles"];
        data = Import["mpg.dat"];
        Dimensions[data]
        dayVals = data[[All, 1]];
        mpgVals = data[[All, 2]];
```

Find the mean and standard deviation

```
In[8]:= Mean[mpgVals]
        StandardDeviation[mpgVals]
```

Plot the data points

```
In[10]:= ListPlot[data, PlotRange → {{0, 2000}, {20, 40}},
          AxesLabel → {"Days since July 1, 2008", "MPG"}]
```

Extract the summer and winter data segments

```
In[11]:= mpgJulAug = Select[data,
              Mod[#[[1]], 365] <= 62 &][[All, 2]];
         mpgJanFeb = Select[data,
              Mod[#[[1]], 365] > 184 && Mod[#[[1]], 365] < 245 &][[All, 2]];
```

Histogram and get the means and standard deviations

```
In[13]:= Histogram[mpgJulAug, {1}, PlotRange → {{24, 34}, Automatic},
          PlotLabel → "MPG for July and August"]
         Mean[mpgJulAug]
         StandardDeviation[mpgJulAug]
```

```
In[16]:= Histogram[mpgJanFeb, {1}, PlotRange → {{24, 34}, Automatic},
          PlotLabel → "MPG for January and February"]
         Mean[mpgJanFeb]
         StandardDeviation[mpgJanFeb]
```

"DataFiles" of the directory in which the notebook is executing. The "Dimensions" command returns "{177, 2}", indicating that the file is 177 lines with two data values on each line. This is a handy format for this kind of data, where the second value is essentially a function of the first value.

The last two lines of the first cell use "Part", in its shorthand form, to separate the first column of data into "dayVals" and the second column into "mpgVals". These are clearly the days and gas mileage readings, respectively.

Next, we find the mean and standard deviation of the mileage values, regardless of the date and time. The values returned are "30.2689" and "2.00811", respectively. In other words, we might quote the average gas mileage of this vehicle as 30.3 ± 2.0 miles per gallon.

The "ListPlot" command produces the following plot:

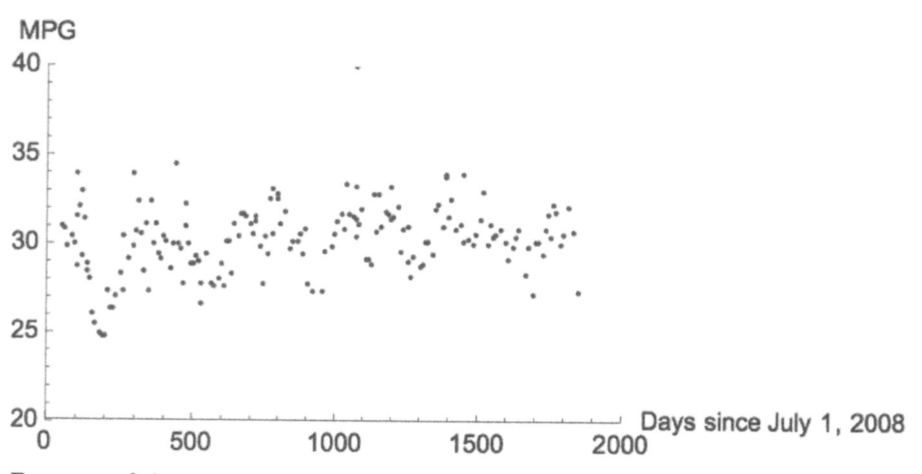

Because of the way "data" is structured, it is simple to plot the second column against the first. I chose "PlotRange" to focus on a specific vertical region, and also to have a neat start and end to the horizontal axis. Data was collected on this vehicle for nearly 2000 days, more than five years. There do appear to be some periodic fluctuations in the gas mileage.

To study this further, the next step is to extract the mileage values corresponding to summer and winter months. The next cell accomplishes this using "Select", but let's look carefully at the selection criteria. Both the summer and winter commands use a pure function operating on the first of "data", namely the number of days since July 1st, and use "Part", once again in the double square bracket shorthand, to select the gas mileage for all of the selected days.

The selection criteria use the "Mod" function, which takes two arguments. The function divides the first argument by the second, and returns the remainder. So, the summer selection asks that the remainder be two months worth of days out of every 365. (If we wanted to be more precise, we could have used 365.25 days.) For the winter months, we require the remainder to be more than half a year, and also less than two months past that.

So, the selection cell looks a little mysterious, partly because of all the "Part" shorthand, but when you parse it through it is actually quite simple.

Finally, we make histograms of the two selected lists of summer and winter mileage. The "Histogram" command uses the same "PlotRange" for each, so the two are easy to compare. The output plots are

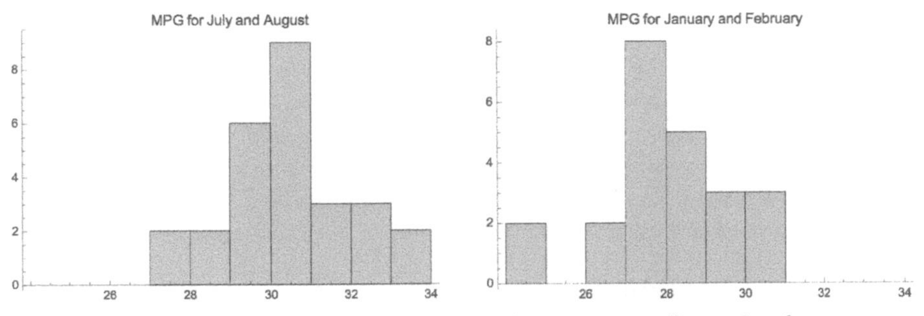

Indeed, it does appear that this car gets better gas mileage in the summer than in the winter. This is confirmed from the calculation of the mean and standard deviation for the two data samples. We find that the mileage in summer is 30.4 ± 1.5 miles per gallon, and in the winter it is 28.0 ± 1.6.

You may want to experiment with this data, looking for differences between early and late days. It so happens that close to day #500, this car had the thermostat replaced, better controlling the flow of warm water to the engine when it needed it.

7.6 CHAPTER SUMMARY

- In its simplest form, "data" is just a list of numbers.
- Basic list operations allow you to combine and manipulate lists of numbers, just as you would for any list.
- There are some special functions for analyzing data in a list, for example "Sum", "Mean", and "StandardDeviation".
- The function "ListPlot" is a simple way to plot lists of data.
- You can make "cuts" on the data using the "Select" function.
- It is possible to read data from a file using the function "Import". You need to be careful about default directories in order to find the file.
- Data can be displayed in histograms. You can also use "Histogram" to get a frequency distribution of a set of data values.

EXERCISES

Data files for selected exercises are available through the publisher's website at the Downloads / Updates tab at https://www.crcpress.com/9781138035096.

7.1 *The file* "Pendulum.dat" *is a two-column list giving the results of a "kitchen" experiment to measure the acceleration g due to gravity. A length of thread was suspended from the ceiling, with a small weight attached to the bottom. After measuring the length, ten swings were timed with a stopwatch. Then the thread was shortened and the process repeated. The file gives the length in inches and the time for ten swings in seconds, for each of the seven lengths. Find the value of g in m/sec^2 for each of the measurements, average them to determine $\bar{g}$ and find the standard deviation σ_g. Does $\bar{g} \pm \sigma_g$ agree with the accepted value?*

7.2 *The file* "ExamGrades.dat" *is a single list of final exam grades in a certain course. For this list of numbers,*

a. *Find the minimum and maximum grades. Find the average grade, and express it is a floating point (decimal) number.*

b. *Show that the first and last elements of the grade list, sorted into increasing order, agree with the minimum and maximum values.*

c. *Use* "Show" *to combine a* "ListPlot" *of the grades with a* "Plot" *of a horizontal line at the average value.*

d. *Make a histogram of the grade distribution, using a bin width equal to 5 and ranging from zero to 100.*

7.3 *The files* "TavgAlbanyNY.csv" *and* "TavgAlbanyTX.csv" *list the average monthly temperatures in Albany, New York and Albany, Texas over more than a century. (The data is downloaded from the Berkeley Earth Project at* http://berkeleyearth.org *using the "US Historical Climatology Network" dataset.) Each file is a two-column list of the date (by year with fraction for the month) and the average temperature (in Celsius degrees).*

Use "ListPlot" *to plot both sets of data on the same graph over several years, to demonstrate the seasonal variation, and the fact that Texas is a hotter climate than New York. The* "Joined" *option will be useful to follow the two sets separately. Include a legend to distinguish the data sets.*

Can you convert the temperatures to Fahrenheit before plotting them?

7.4 *Use the data from Exercise 7.3 to study the average temperature over the periods 1900–1930, 1940–1970, and 1980–2010. Choose either Texas or New York, and one particular month. (Examine the data set and you'll see how the month is specified in the "year" value. You might want to learn the function* "FractionalPart" *to select the month.) Find the mean value for each of the 30 year periods, and use* "SmoothHistogram" *to histogram each period on the* *same graph. Include a legend, as well as a plot label that indicates your choice of state and month.*

7.5 *So-called "Cepheid Variable" stars are useful distance markers because they pulse with a period directly related to their intrinsic brightness. The distance to the galaxy M81 was determined by Freedman, et al., The Astrophysical Journal 427 (1994)628, using this technique. The paper lists the brightness (as "magnitude") on 31 stars taken at various times over more than a year, and also the periods of each of these as determined from their analysis.*

Choose three stars with rather different periods, and plot the magnitude versus time over one period. (You can download the data from the file "m81Cepheids.csv" *instead of typing it in yourself.) You'll need to use the function* "Mod" *to fold all the days into one period. Embellish the plots how you see fit. Note that, in astronomical terms, a brighter star has a smaller magnitude value, so you should try plotting the vertical axis in reverse order, as done in the paper. An option like* "ScalingFunctions → {None, "Reverse"}" *in* "ListPlot" *will do this for you.*

7.6 *The file* "MilliCan.dat" *contains a list of 12 numbers, each a measurement of the mass (in grams) of a sealed container, each holding some number of identical ball bearings. Use this data to find the mass of a single ball bearing, and estimate the uncertainty in this mass.*

CHAPTER 8

Fitting Data to Models

CONTENTS

When analyzing data, a physicist will often want to compare trends in the data with some model or set of models. That model is generally expressed as a mathematical function. This chapter discusses how to fit functions (or "curves") to data, and to test how well that fit agrees with the data.

It is important to realize that whatever is learned from this procedure is "model dependent," at least to some extent. The model will be described by some mathematical function (or set of functions) which has some freedom in its detailed description. This freedom appears as "free parameters" which are adjusted to give the "best fit" to the data.

If the model is based on some practical physical theory, which should be but is not always the case, then the fitted values for the parameters can be compared to a theoretical prediction. The extent to which this comparison agrees (or not) with theory is generally the main purpose of the exercise.

In order to make this comparison quantitative, one needs to take into account the uncertainties, or "errors", in the data. That is, each of the data points included in the fit has some uncertainty in its value, and these uncertainties may certainly be different for different data points. MATHEMATICA provides the ability to include these uncertainties in the fit procedure, and to plot the data with "error bars" that display the uncertainties on each point.

All fitting algorithms work to adjust the fit parameters to minimize the difference between the fit function and the data values. When the fitted function depends only linearly on the fit parameters, an analytic algorithm can be used to determine the best fit parameters. If the dependence is nonlinear, a numerical algorithm is used. MATHEMATICA has different functions for these two cases, as well as a more extensive package for dealing with data that has uncertainties.

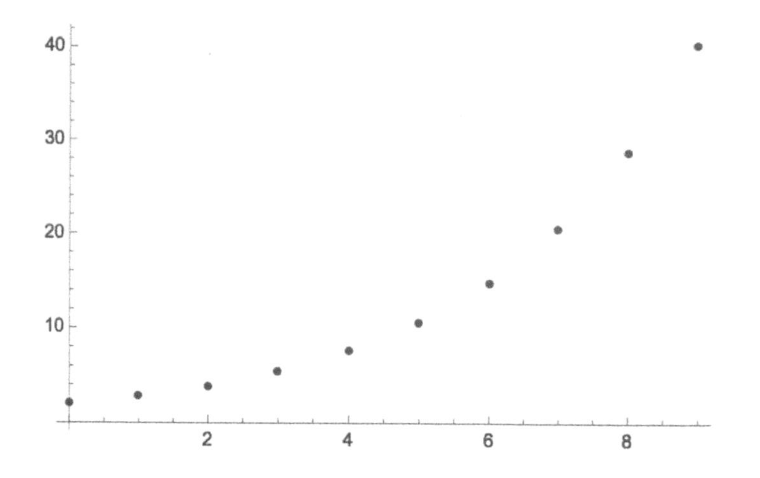

FIGURE 8.1 This plots the simple "data" set used to demonstrate fitting techniques in Sections 8.1 and 8.2, generated by the function $2e^{x/3}$.

8.1 LINEAR FITTING

Let's start by giving ourselves some "data" to fit, generated according to a known function. We can try out various functions and compare them to the data, and then we can try the actual function and see if the fit gives us the same thing we started with.

We can easily generate a data set using "Table". (Recall Section 6.4.) The following cell is a systematic way to create a dataset that will serve our purposes here:

```
func = a Exp[x/b];
vals = {a -> 2, b -> 3};
data = Table[{x, func}, {x, 0, 9}] /. vals;
```

The object "data" is simply ten pairs of points, the first of the pair ("x") being the integers from zero to nine, and the second the value $2e^{x/3}$. It is helpful to visualize the data, and "ListPlot[data]" produces the plot shown in Figure 8.1.

What does it mean to "fit a curve" to data? The "curve" is a functional form with some number of parameters whose values are determined by making the curve agree as closely as possible with the data points. There are a number of techniques used, and criteria employed, to determine what is the "best fit", but we will not get into that in any detail in this book. On the other hand, MATHEMATICA provides us with a number of standard ways to reach this goal.

The simplest type of fit uses a functional form that is linear in the free parameters. That is, the fit function has the form

$$f(x) = a_1 g_1(x) + a_2 g_2(x) + \cdots + a_n g_n(x)$$

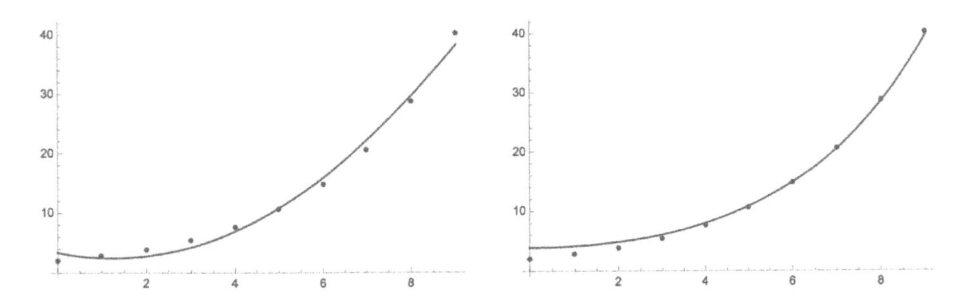

FIGURE 8.2 Fits to the "data" in Figure 8.1 to two different linear models. See the text for details.

where the $g_i(x)$ are functions chosen on the basis of the theoretical model, and the a_i are free parameters to be adjusted to fit the data. This kind of parameterization is particularly well suited to fitting based on the "method of least squares", where the a_i are chosen to minimize the sum of the squares of the differences between the model and the data points. For a linear fit, this algorithm reduces to a simple problem in matrix algebra.

The MATHEMATICA function used to perform a linear fit is "Fit". It takes three arguments. The first is the data, as a nested list of pairs of points. The second is the list of functions (expressions) $g_i(x)$ that make up the model, and the third is the independent variable, in this case x. The function returns the expression with the best fit paramters.

So, for example, to fit the data plotted in Figure 8.1 to a parabola, execute

```
fit1 = Fit[data, {1, x, x^2}, x]
```

which returns "3.4026 − 1.51708x + 0.599649x^2".

It is natural to want to compare the fit to the data. Figure 8.2 shows, in the left panel, this comparison. Indeed, the curve comes close to the points, sometimes missing above, and sometimes missing below, but seems to be a reasonable "fit" to the data.

The plot was made by combining the outputs of "ListPlot" and "Plot" with the function "Show", and executing the cell

```
Show[
 ListPlot[data],
 Plot[fit1, {x, 0, 9}]]
```

That is, "Show" simply combines the outputs of the two plotting commands. We'll make more use of "Show" later.

The function "fit1" does a pretty good job of approximating the data, but we could do better. Certainly we could just add an x^3 term, then an x^4 term, and so on. However, the expressions combined to make our linear fit do not have to be simply powers of x. We could try $\cosh(x/3)$, for example. We know this might do a better job all by itself, because $\cosh(z) = [e^z + e^{-z}]/2$, so it is close to the actual function we used to generate the data. Indeed

```
fit2 = Fit[data, Cosh[x/3], x]
```

returns an expression that is plotted in the right panel of Figure 8.2. The agreement with the data does in fact appear to be better.

Of course, we would prefer to have a quantitative comparison of the two fits. Since the fitting algorithm aims to minimize the sum of the squares of the differences between the data and the fit, it makes sense to compare this number for the two fits. Although MATHEMATICA offers some sophisticated ways to make this comparison, let's work through it here from the basics.

To get this value, generically called χ^2, for the polynomial fit, execute

```
fitVals = fit1 /. x -> data[[All, 1]]
sqdiffs = (fitVals - data[[All, 2]])^2
chisqur = Total[sqdiffs]
```

These commands should all be familiar to you by now, except perhaps for "Total" which simply adds up all the members of the list. A value 12.9987 is returned for "chisqur". This is our measure of how good an approximate "fit1" is to the data.

Repeating the same procedure for "fit2" returns the value 7.52662 for "chisqur". Indeed, the fit using $\cosh(x/3)$ is better than using a parabola.

Before moving on to nonlinear fitting, there is one more thing to cover. We have seen how to find the linear functional form that best fits a set of data, but suppose it is the fit coefficients that we're after. The easiest way to get these from the fit function is to use the functions "Coefficient" or "CoefficientList". These simply extract the coefficients from the fitted expression. For example, referring to the result "fit1" above, executing

```
Coefficient[fit1, x]
```

returns the value "-1.51708". Executing

```
CoefficientList[fit1, x]
```

returns the list "{3.4026, -1.51708, 0.599649}". For the result "fit2", execute

```
Coefficient[fit2, Cosh[x/3]]
```

to get "3.94329".

8.2 NONLINEAR FITTING

Of course, we'd like to fit the data from the last section with the actual function that we used to generate it. The expression $ae^{x/b}$, however, is a nonlinear function of the parameters, so we cannot use "Fit" to determine a and b.

Nonlinear fits are performed with “FindFit”. The goal is again to minimize the sum of squares of the deviations, but now the algorithms are numerical instead of analytic. The details are not important here, but “FindFit” is nevertheless easy to use.

To perform a fit to our generated data with the “true” functional form, we can execute the cell

```
fitPars = FindFit[data, func, {a, b}, x]
fitTrue = func /. fitPars;
```

which returns the replacement list “{a → 2., b → 3.}”. You can then determine the χ^2 using

```
fitTrue = func /. fitPars;
fitVals = fitTrue /. x -> data[[All, 1]]
sqdiffs = (fitVals - data[[All, 2]])^2
chisqur = Total[sqdiffs]
```

and discover that “chisqur” returns 2.82856×10^{-28}. It is not zero, due to roundoff error, but obviously very tiny.

Let’s try something a little more interesting, and fit a Gaussian shape to the histogram we created in Section 7.4. You’ll recall that this histogram came from binning a data set that was generated according to a Gaussian distribution. Fitting the histogram to a Gaussian should give us back the same parameters used to generate the data in the first place.

First let’s review the steps that get us the binned histogram data:

```
SetDirectory[NotebookDirectory[]];
SetDirectory["./DataFiles"];
rg = Flatten[Import["rangs.dat"]];
bList = {-5, 8, 0.25};
gData = HistogramList[rg, bList];
```

Now we have to be a little careful. We want the data arranged as a list of $(x.y)$ pairs, but that’s not what “HistogramList” returns. Instead, it gives us one list of “bin edges”, and one list of histogram “heights”, combined into one nested list. These two lists have different lengths, because there is one more “bin edge”, namely the high side of the highest bin, than there are “heights”. Indeed “Length[gData[[1]]]” returns “53”, and “Length[gData[[2]]]” returns “52”.

So let’s do the following. We can remove the last element of the bin edge list using the Mathematica function “Drop”, and then build an appropriate nested list that we can transpose. While we’re at it, we’ll change the first element of our data pair so that it is the middle of a bin instead of the left edge. (This makes more sense from the point of view of the fitting exercise.)

The commands that carry this out are the following:

```
xVals = Drop[gData[[1]], -1] + bList[[3]]/2;
yVals = gData[[2]];
xyPairs = Transpose[{xVals, yVals}];
```

Now we can carry out the Gaussian fit. Execute the cell

```
gFunc = a Exp[-(x - mean)^2/(2 sigma^2)];
gPars = FindFit[xyPairs, gFunc, {a, mean, sigma}, x]
```

to find that

```
{a -> 39.1918, mean -> 1.64179, sigma -> 2.55316}
```

The result can be plotted using

```
gFit = gFunc /. gPars;
Show[
 ListPlot[xyPairs],
 Plot[gFit, {x, -5, 8}]]
```

which returns the plot shown here:

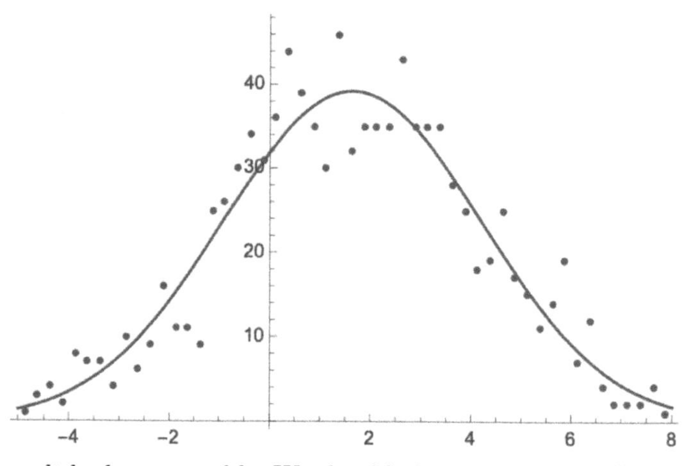

This result looks reasonable. We should also note that the fitted mean and standard deviation are reasonably close to the values 1.5 and 2.5, respectively, that we used to generate the original data distribution.

8.3 HANDLING DATA WITH ERROR BARS

Everything we've done so far in this chapter assumes that each data point carries equal weight. However, this is rarely the case in most physics experiments. One data point will carry more or less weight than another because of random statistical fluctuations, systematic uncertainties, or both.

MATHEMATICA provides a number of different functions aimed at fitting data in ways much more general than what we have covered so far in this chapter. There is a tutorial on *Statistical Model Analysis* which covers all of this, and this section will just cover some of the simplest examples. You should also review the language guide on *Scientific Data Analysis* which includes much of the material discussed in this section.

Let's first generate[1] some data to use as an example, and in the process show how to make plots of data points with error bars. First, executing the cell

```
nPts = 21;
xVals = Table[i - nPts/2 - 1/2, {i, nPts}];
```

gives us a list of 21 x values that are the integers between -10 and $+10$. Define a linear model using

```
m = 3;
b = -1;
uVals = m xVals + b;
```

Now we can add some scatter to the points with

```
dSize = (Max[uVals] - Min[uVals])/4;
dVals = RandomReal[{-dSize, dSize}, nPts];
yVals = uVals + dVals;
```

The function "RandomReal" is used to randomly add or subtract a value to the model function. In other words,

```
xyVals = Transpose[{xVals, yVals}];
```

creates a list of pairs of points, that more or less fall along our model. We could plot these, as we have already seen, using "ListPlot[xyVals]", but let's move on and generate some uncertainties for the data points. This is simple enough with a statement like

```
eVals = RandomReal[{dSize/4, 3 dSize/4}, nPts];
```

Now we have a set of "data" that we understand as a point with uncertainty $(x, y \pm e)$ with these data contained in the lists "xVals", "yVals", and "eVals", respectively.

In order to plot this data set, we need to do something we have not yet encountered. MATHEMATICA does not automatically load all of the "packages" that it might need, and making plots with error bars is one of those. You can load the appropriate package manually with

```
Needs["ErrorBarPlots`"]
```

---

[1]We will use simple random number generation, which will actually be covered more thoroughly in Section 10.1.

(There are ways to have MATHEMATICA load this, and other, packages as defaults, but we won't get into that here.) This package makes available the function "ErrorListPlot" which creates plots with error bars. For our case, in its simplest form, you could use "ErrorListPlot[Transpose[yVals, eVals]]", but this will plot the index of the list on the horizontal axis, and not the x-values.

In order to make the plot we want, we need to get into some of the many options for "ErrorListPlot". The data points $(x, y \pm e)$ are input to "ErrorListPlot" as a list whose elements are of the form

```
{{x,y},ErrorBar[e]}
```

where "Errorbar" is also a MATHEMATICA function that specifies details of the error bar. To build this list, execute the cell

```
eBars = Table[ErrorBar[eVals[[i]]], {i, 1, nPts}];
xyeVals = Partition[Riffle[xyVals, eBars], 2]
```

The first statement is straightforward, but note the use of the shorthand for "Part". The second statement first uses "Riffle" to interleave the "{x,y}" pairs with the "ErrorBar[e]", and next uses "Partition", to group every two elements of the interleaved list, giving us, finally, a list with elements in the form "{{x,y,ErrorBar[e]}}". You'll notice that I left the semicolon off the end of the second statement. It is a good idea to examine the list "xyeVals" to make sure it has the correct structure.

We can now execute

```
ErrorListPlot[xyeVals]
```

to get the following plot:

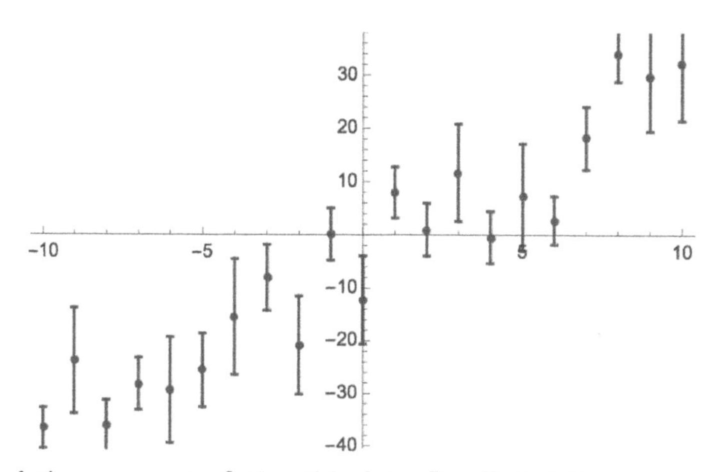

Now let's move on to fitting this data. Recall that it was generated according to the model $y = 3x - 1$. Indeed, "Fit[Transpose[xVals, uVals], 1, x, x]" returns exactly this result. Fitting the data with the random scatter on the y values using

```
Fit[xyVals, {1, x}, x]
```

returns

```
-4.41115 + 3.35366 x
```

Now, however, we want to fit including the error bars.

The proper form of the χ^2 function for data with uncertainties is

$$\chi^2 = \sum_i \frac{[(y_i - f(x_i)]^2}{e_i^2}$$

for a model $y = f(x)$ and where the uncertainty (error bar) on a point (x_i, y_i) is e_i. The factors $w_i = 1/e_i^2$ are "weights" for the individual data points.

MATHEMATICA provides two functions for carrying out weighted fits. These are "LinearModelFit" and "NonlinearModelFit" with the obvious different applications. Using these in their simplest form with no options is equivalent to "Fit" and "FindFit", discussed in Sections 8.1 and 8.2, respectively.

Let's first use the more sophisticated "LinearModelFit" with the same, simple straight line model we've been using, and with no weights. We should get the same result as with "Fit". The syntax is

```
LinearModelFit[xyVals, x, x];
Normal[%]
```

and this cell returns

```
-4.41115 + 3.35366 x
```

just as we expect. You should remove the semicolon and see the result returned by "LinearModelFit". This will give you a glimpse into the extended capabilities of this fitting package. Here, however, we use "Normal", which we have encountered before (See page 35) to convert the output into the form we are used to.

We apply weights to the fit by using"LinearModelFit" with the "Weights" option. Executing the cell

```
wVals = 1/eVals^2;
fitw = LinearModelFit[xyVals, x, x,
   Weights -> wVals];
Normal[fitw]
```

returns

```
-4.59475 + 3.28697 x
```

and you can see that the result is slightly different. We can make a plot of the data, the input model, and the fit result, with

```
Show[
 ErrorListPlot[xyeVals],
 Plot[m x + b, {x, Min[xVals], Max[xVals]},
  PlotStyle -> {Dashed, Red}],
 Plot[Normal[fitw], {x, Min[xVals], Max[xVals]},
  PlotStyle -> Blue]]
```

which produces the plot

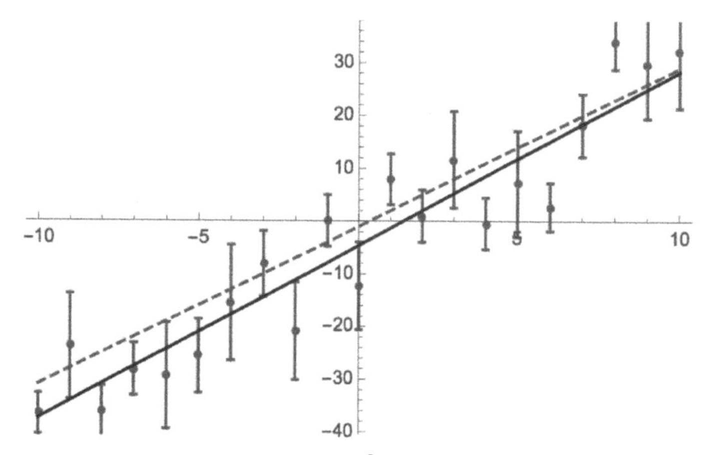

It is instructive to calculate the χ^2 for the fit. You can do that easily enough by executing

```
chisqr = Total[
  (yVals - (Normal[fitw] /. x -> xVals))^2/eVals^2]
```

which returns the value 29.8. With 21 data points and 2 free parameters, we say that this problem has 19 "degrees of freedom." A rule of thumb for statistical analysis is that the (weighted) χ^2 should have a value somewhere near one per degree of freedom. It would seem that this is a reasonable fit to the data, but of course, I chose values for the random number generator that would make it come out that way.

So far, I have been referring to "LinearModelFit" (or "NonlinearModelFit") as just another function that fits data to a model, albeit with the ability to includes weights. Indeed, they are much more than that. The tutorial on *Statistical Model Analysis* goes through a lot of this, but I will give you some pointers here on how to make use of the available features.

A statement like

```
fitw = LinearModelFit[xyVals, x, x,Weights -> wVals];
```

actually defines a function "fitw" which can be used in various ways, much more than just input to "Normal". Executing the cell

```
fitw["Properties"]
```

returns a long list of various quantities and information that can be accessed with "fitw". For example,

```
fitw["BestFitParameters"]
```

and

```
fitw["ParameterErrors"]
```

return the fit parameter values and errors, respectively. Executing

```
diff = fitw["FitResiduals"];
```

returns a list of the differences between the fit and data. There are many more options, and you should study the *Statistical Model Analysis* tutorial for more information. We will make use of these features in Example 8.1.

8.4 PHYSICS EXAMPLE

Example 8.1 *The file "Cs137.dat" contains the result of a measurement of the γ decay of $^{137m}Ba^*$, following the long lived β decay of ^{137}Cs. The data is in two columns representing "time" and "counts," with time measured in 20-second intervals. This state is known to decay wilh a half-life of 2.552 minutes, but there is a constant background due to residual cesium after the chemical separation to extract the barium. Analyze this data and confirm the half-life.*

Let's take a moment to review the basic experimental physics notions needed to do this exercise. First, the number of counts $N(t)$ as a function of time should follow the equation

$$N(t) = N_0 2^{-t/t_{1/2}} + b \tag{8.1}$$

where N_0 is the initial rate, $t_{1/2}$ is the half-life, and b is the constant background term. If $b \ll N_0$, then at early times we expect that

$$\ln N(t) = \ln N_0 - \left(\frac{\ln 2}{t_{1/2}}\right) t$$

so the logarithm should follow a straight line in the beginning. Finally, you also need to understand that for radioactive decay data measuring "counts", the uncertainty in each point follows a Poisson distribution, in which case the standard deviation on N is $\sqrt{N}$.

Now we can apply MATHEMATICA to analyze this data. Notebook 8.1 shows an approach using unweighted nonlinear fitting. (I have once again suppressed the output to save space in the figure.)

The first section of the notebook is just what we have seen before, reading in the data file and extracting the columns into their own lists. A new list

NOTEBOOK 8.1 One possible solution using unweighted fits for Example 8.1.

```
In[1]:= Remove["Global`*"]
```

Radioative decay of 137Cs

Read the file and prepare the data and errors

```
In[2]:= SetDirectory[NotebookDirectory[]];
        SetDirectory["./DataFiles"];
        dataFile = Import["Cs137.dat"];
        timeTics = dataFile[[All, 1]];
        cntsVals = dataFile[[All, 2]];
        erroVals = Sqrt[cntsVals];
```

Convert time to minutes and plot the data

```
In[8]:= timeVals = timeTics / 3;
        data = Transpose[{timeVals, cntsVals}];
        ListLogPlot[data,
         AxesLabel → {"Time (Mins)", Counts}]
```

Unweighted fits, with tHalf floating or fixed

```
In[11]:= funcFull = a 2^(-t / tHalf) + b;
         parsFull = FindFit[data, funcFull, {a, tHalf, b}, t]
         funcFixT = funcFull /. tHalf → 2.552;
         parsFixT = FindFit[data, funcFixT, {a, b}, t]

In[15]:= Show[
          ListPlot[data,
           PlotRange → {0, a /. parsFull}],
          Plot[funcFull /. parsFull, {t, 0, Max[timeVals]},
           PlotRange → {0, a /. parsFull}]]

In[16]:= rFloat2Fixed =
           (funcFull /. parsFull) / (funcFixT /. parsFixT);
         Plot[rFloat2Fixed, {t, 0, Max[timeVals]}]
```

"erroVals" is created to define the error bar values as the square root of the number of counts.

Next we plot the data. A list "timeVals" is defined which is just the time bins converted to minutes. Then, as we have done before, a two-dimensional list is created for input to the plot function. This time, however, instead of "ListPlot", we use "ListLogPlot" which plots the vertical scale on a logarithmic axis. (The syntax and options are same for "ListPlot" and "ListLogPlot".) The output is below. Notice that, on a log plot, it is easy to see the exponential decay behavior of the data; at low times it falls linearly, as we expect.

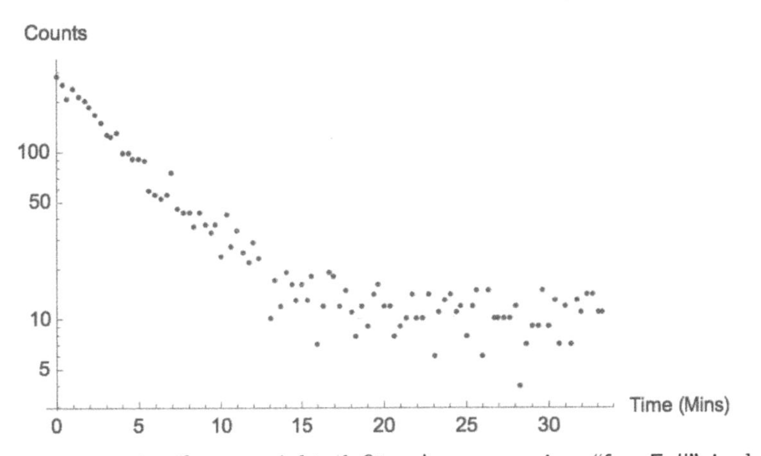

Now move on to the unweighted fits. An expression "funcFull" is defined to be the radioactive decay plus a constant background. Note that this is nonlinear in the fit parameter "tHalf" for the half-life, so we need to use "FindFit" for the best fit parameters. It returns

```
{a -> 272.064, tHalf -> 2.71357, b -> 9.90075}
```

indicating a half life of 2.714 minutes, close to the known value 2.552 minutes. In order to get some idea of how close, we try fitting with the half-life fixed to the known value by defining the expression "funcFixT". Even though this is linear in the two remaining parameters, we still use "FindFit" just for consistency. In this case, we find

```
{a -> 277.215, b -> 11.1862}
```

These values are very close to those we find in the three-parameter fit.

We'll continue this comparison now using plots. The next cell in Notebook 8.1 uses "Show" to combine a "ListPlot" of the data with a "Plot" of the full function with the parameters replaced by the result of the full fit:

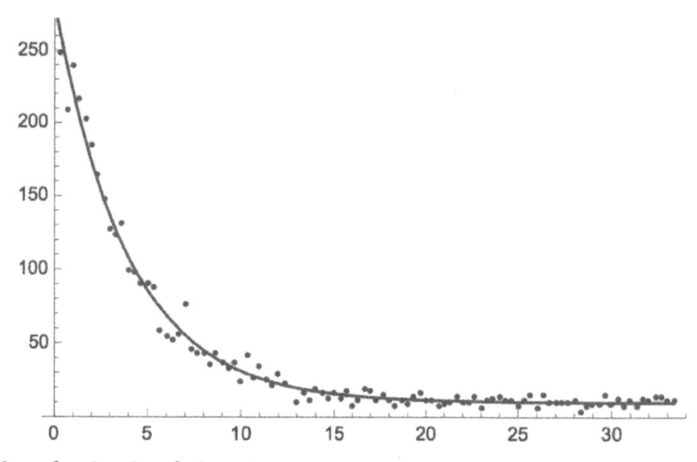

Notice that for both of the plot commands, I specify a vertical range based on the fitted number of counts at zero time. The fit appears to be quite good, with the curve going neatly through the points.

We could compare the fixed half-time fit by including a third plot using "funcFixT", but it will be hard to see much of a difference with the full fit. (You should try it and see.) Instead, the notebook defines "rFloat2Fixed" to be the ratio of the expression from the full fit to that with the fixed half-life. The output of the "Plot" command is shown here:

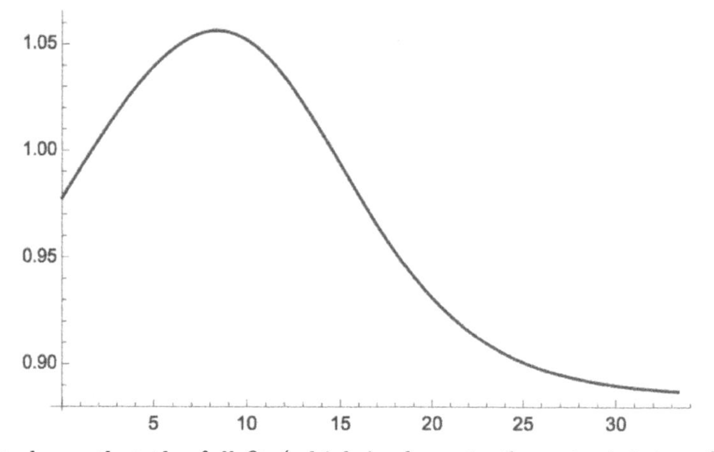

The plot shows that the full fit (which is closer to the actual data, of course) is about 5% higher at low times, and down to 10% lower at long times.

So, the fit looks good, and the fitted half life is "close" to the actual one, but in order to be more quantitive about the uncertainties on the parameters and the goodness of fit, we need to use "NonlinearModelFit" which is done in the next section of the notebook. This next section is shown in Notebook 8.2, which continues on from Notebook 8.1.

The notebook uses "NonlinearModelFit" with the appropriate "Weights" option and stores the result in the object "efit. Remember that this is actually

NOTEBOOK 8.2 A weighted model fit for Example 8.1. This is a continuation of the notebook shown in Notebook 8.1.

Weighted fit with floating half life parameter

```
In[18]:= efit = NonlinearModelFit[data, funcFull,
            {a, tHalf, b}, t,
            Weights → 1/erroVals^2];

In[19]:= Needs["ErrorBarPlots`"]
         eBars = Table[ErrorBar[erroVals[[i]]],
            {i, 1, Length[timeTics]}];
         edata = Partition[Riffle[data, eBars], 2];
         Show[
          ErrorListPlot[edata],
          Plot[Normal[efit], {t, 0, 30}]]
```

Learn about the fit and parameters

```
In[23]:= efit["BestFitParameters"]
         efit["ParameterErrors"]

Out[23]= {a → 275.517, tHalf → 2.63599, b → 9.36367}

Out[24]= {7.6324, 0.0736614, 0.483013}

In[25]:= diffData = Partition[
            Riffle[
             Transpose[{timeVals, efit["FitResiduals"]}],
             eBars], 2];
         ErrorListPlot[diffData]

In[27]:= Total[efit["StandardizedResiduals"]^2]/
          (Length[timeVals] - 3)
```

a function which can be used in a number of ways. Before we examine the parameters, let's take a look at how the fit compares to the data.

The next cell loads the ErrorBarPlots package and prepares to make a plot with error bars, overlaid with the results stored in "efit", just as we did in Section 8.3. That is, a list is created using "Table" of the function "ErrorBar" of the error values, and then this list is merged with the data in accord with the syntax for "ErrorListPlot". We then use "Show" to combine the plot of data with error bars, and the "Normal" form of the result from "NonlinearModelFit":

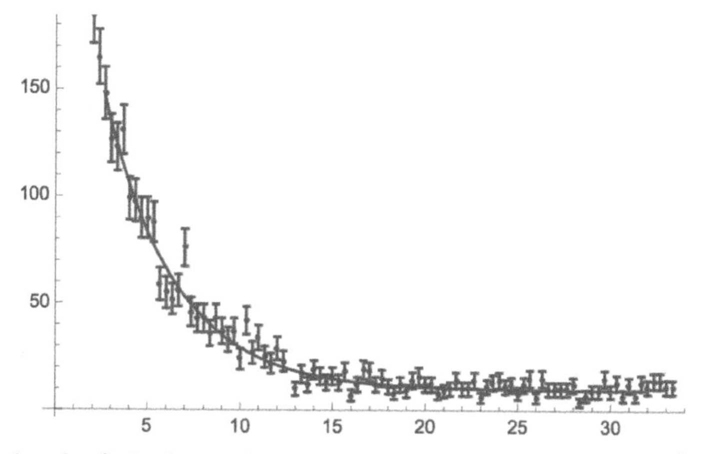

Once again, the fit looks good, and the error bars look reasonable.[2]

The last section of the notebook is used to, finally, get detail on the fitted parameters and the goodness of the fit. The command "efit["BestFitParameters"]" returns a replacement list for the three parameters, and the command "efit["ParameterErrors"]" gives the uncertainties on those parameters. This time, I've included the output of these commands. Resorting to the notation in Equation 8.1, and using only appropriate significant figures, the result is

$$\begin{aligned} N_0 &= 275.5 \pm 7.6 \\ t_{1/2} &= 2.636 \pm 0.074 \text{ minutes} \\ b &= 9.36 \pm 0.48 \end{aligned}$$

Now we see that the half life is in good agreement with the accepted value, within uncertainties. (Actually, it is about one standard deviation away from the accepted value. Perhaps the student who took this data should investigate for potential sources of systematic uncertainty.)

The notebook investigates the goodness of fit, both graphically and by calculating the χ^2. Another property FitResiduals is used with "efit" to get the difference between the fit and the data points. We use, again, the same

[2]A rule of thumb for error bars is that every third data point should miss the curve by one or more standard deviations.

combination of the functions “Transpose”, “Riffle”, and “Partition” to come up with a data structure that allows us to use “ErrorListPlot” to plot these residuals with errorbars:

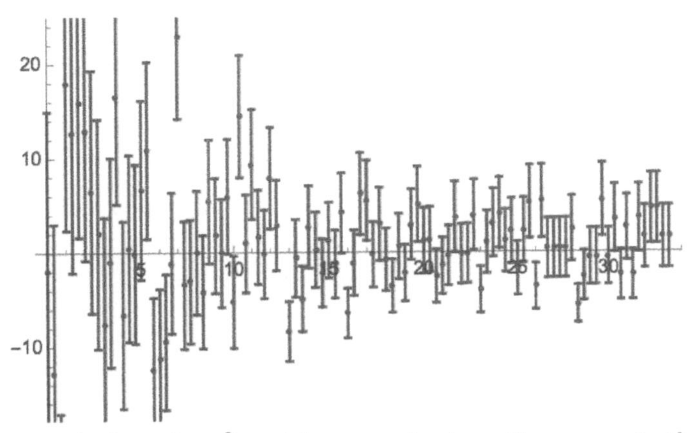

Finally, I calculate the χ^2, making use of yet another property that can be returned by “efit”, namely StandardizedResiduals. This list is just the difference between the fit and the data, divided by the uncertainty on each point. (Plotting this quantity, instead of the FitResiduals with error bars, is also a handy way to graphically investigate the goodness of fit.) Using “Total” to sum the squares of the StandardizedResiduals and then dividing by the number of data points minus the number of free parameters, we find that $\chi^2 = 1.034$ per degree of freedom. This is indeed a good fit to the data.

8.5 CHAPTER SUMMARY

- Using “Fit” it is straightforward to fit a data set to a simple linear model.
- Nonlinear models can be fit to data using “FindFit”.
- It is more complicated to include weights in the fit and in plots with “error bars”. For these, you should resort to the functions “LinearModelFit” and “NonlinearModelFit” and become familiar with their properties.

EXERCISES

Data files for selected exercises are available from the publisher’s website under the Downloads / Updates tab at https://www.crcpress.com/9781138035096.

8.1 *The ideal gas law relates the pressure P, volume V, and temperature T for a gas made of n particles as $PV = nkT$. Clearly $T = 0$ implies that $P = 0$, but any gas becomes far from ideal long before the temperature is reduced to anything close to this “absolute zero” value. However, we can measure the value T_0 of absolute zero by using the Celsius temperature scale, in which case $T = T_0 + T_C$. That is, where the temperature T_C is measured in Celsius degrees.*

The file "idealgas.dat" is two columns of data, measuring the pressure of a fixed, closed volume of gas, for different temperatures. The first column is the Celsius temperature and the second is the pressure in Pascals. Use this data to determine absolute zero T_0 on the Celsius temperature scale. Make a plot of your fit, extending the temperature scale low enough to see where the pressure would go to zero. Compare your value of T_0 to the accepted value.

8.2 *In Compton scattering, a photon of incident energy E scatters from an electron with mass m through an angle θ relative to the incident direction. The scattered photon energy is*

$$E' = \frac{E}{1 + (E/mc^2)(1 - \cos\theta)}$$

The file "Compton.dat" contains data for a Compton scattering measurement for a specific incident energy E. The first column is the angle θ in degrees, and the second is E' in MeV. Fit $1/E'$ versus $1 - \cos\theta$ to a straight line, and determine the incident photon energy and electron mass (mc^2).

8.3 *Use the data from Exercise 7.1 to once again determine g, but this time use the functional form $\tau = 2\pi\sqrt{\ell/g}$ for the period τ of a pendulum of length ℓ. Do this both using a linear fit of τ^2 versus ℓ, and then using a nonlinear fit to the original functional form. In both cases, make a plot of the data with the fit superimposed.*

8.4 *The force exerted by a physical spring might be written as $F = -kx + \alpha x^2$ where x is the spring extension, k is the spring constant, and the term αx^2, where $\alpha > 0$, represents a weakening of the spring if x gets too large. Data was taken to test this relation using the following apparatus:*

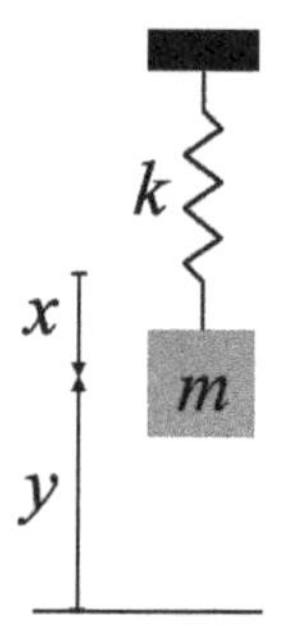

Different masses m are hung from the bottom of the spring. For each mass, the height y above the floor is recorded. Clearly, $x = y_0 - y$ for some constant y_0.

The file "SpringData.dat" contains the results of six measurements of m in kilograms and y in meters, in a two column format. Determine a function $y = y(m)$ in terms of k, α, and y_0. First perform a linear fit to the data fixing $\alpha = 0$, note the results for k and y_0, and plot the data with the fit, and also *the residuals, that is, the differences between the six points and the fit. You will observe a systematic deviation of the residuals as a function of mass.*

Second, perform a nonlinear fit allowing α to float. (If you use "FindFit", *you will want to check the documentation to learn how to set starting values for the parameters. You can set k and y_0 to start with the values from the linear fit.) Note the fit parameter values and how they compare to the linear fit. Do k and α make sense? Once again, check the fit by plotting it over the data points, and examine the residuals to see if the systematic deviation goes away.*

8.5 *When natural silver is exposed to a thermal neutron source, two short-lived radioactive isotopes are produced. One has a half life of 24.4 seconds and the other is 2.42 minutes. The data from such a measurement is contained in the data file* "silver.dat"*. The first column has the time after irradiation in digitizer channels, and the second column has the number of decays counted in that time interval. Each time interval is 2.5 seconds.*

Use this data to determine the two half lives and compare the result to the accepted values.

It will be useful to rearrange the data into four-channel groups, so that the more sparse time bins have a finite number of counts. You can do this using "Partition" *on the individual channel and decay counts lists. Following with* "Total" *will add up the counts in the adjacent four channel bins, and* "Mean" *will form the average channel number.*

Make a plot with error bars of the data, the final fitted function, and also the separate "fast" and "slow" decay components of the best fit function. Use different colors for the data points and each of the three functions plotted.

8.6 *The radioactive nucleus ^{60}Co emits two gamma rays for every decay. One gamma ray has energy 1.17 MeV, and the other 1.33 MeV. These gamma rays are efficiently detected by large NaI crystals, albeit with limited resolution. The crystals are connected to photomultipliers, and the electrical signals are digitized to a number proportional to the gamma ray energy.*

The file "Co60.dat" *contains two-column data. The first column is the digitizer value, and the second is the histogram value of the energy from a long exposure to a ^{60}Co source. Read in the file and plot the data. You'll see a large amount of structure, and two very prominent peaks above digitizer channel 5000.*

Select the data between channels 5000 and 8000, and plot this subset, accentuating the two peaks. Fit the data to the sum of two gaussians plus a linear "background" function. (You may need to specify starting values for the eight parameters.) Check the ratio of the peak values against the ratio of gamma ray energies. How well do they agree? Make a plot of the deviation between the fit and the data, and calculate a χ^2. Does the value seem correct to you?

CHAPTER 9

Numerical Manipulations

CONTENTS

MATHEMATICA can be used as a very sophisticated numerical calculator. This includes getting numerical solutions to ordinary algebraic equations and extracting values from vast databases of commonly used numerical values. This chapter shows you how to make use of these capabilities.

Numerical precision is a concept that you need to appreciate, though, because when it comes to numbers, MATHEMATICA can only do as well as the computer it runs on. Numbers are, of course, stored as a series of ones and zeroes, that is binary digits or "bits." For integers (also known as "fixed point numbers"), there are no issues with precision. You can perfectly represent any integer in base two using the right series of bits.

We will see, however, that non-integer, and especially irrational, numbers (also known as "floating point numbers") present a special problem with precision. Somehow, MATHEMATICA has to decide how many decimal places are an accurate representation of the result of a calculation. This chapter will help guide you through some of the functions that provide information on the accuracy of a numerical calculation.

MATHEMATICA makes it possible to carry out numerical calculations with numbers attached to units. For example, you can multiply 10 ft by 10 cm and get the answer in square feet. This chapter will point you in the right direction for learning more about this sort of thing.

Lastly, MATHEMATICA "knows" the values of all the commonly used physical parameters you are likely to need. It even provides those values with units attached, if you like. Furthermore, it will go out and find less commonly used values if you ask it to. If you get used to doing things this way, it will make many calculations much easier to execute, but it's a good idea to always check along the way to make sure you're getting what you think you're getting.

9.1 SIGNIFICANT FIGURES

We first saw the function "N" back in Section 1.2, but it wasn't mentioned there that it has an optional second argument that is the number of significant figures. That is,

```
N[Pi]
```

returns

```
3.14159
```

but

```
N[Pi, 25]
```

returns

```
3.141592653589793238462643
```

This function also works with rational numbers, such as "N[1345/21398, 10]" which returns "0.06285634171". Note that special functions return results in "machine precision", so there is no affect on the number of significant figures. That is, "Sin[0.1]" and "N[Sin[0.1], 10]" both return "0.0998334".

Let's use this for a simple but instructive exercise in mathematics. I suppose you are aware that the base of the natural logarithm is

$$e \equiv \lim_{x\to\infty}\left(1+\frac{1}{x}\right)^x$$

(If you don't know this, I suggest you go through the calculation of the derivative of the function $f(x) = \log_b(x)$ and show that $f'(x) = 1/x$ implies that $b = e$.) The command

```
eTable = Table[{n, 10^n, N[(1 + 1/10^n)^(10^n), 10]}, {n, 0, 8}];
```

creates a nested list of a sequence of approximations to e for $x = 10^n$ where $n = 0, 1, 2, \ldots, 8$. (The calculation takes some time, almost 33 seconds on my computer.) Executing

```
Grid[eTable]
```

outputs the result of this nested list as a two-dimension grid, which I show here making use of "TeXForm" to format the output (and to add headings):

n	$x = 10^n$	$(1+1/x)^x$
0	1	2.000000000
1	10	2.593742460
2	100	2.704813829
3	1000	2.716923932
4	10000	2.718145927
5	100000	2.718268237
6	1000000	2.718280469
7	10000000	2.718281693
8	100000000	2.718281815

This can be compared to the true value of e up to ten significant figures by executing "N[E, 10]" which returns "2.718281828".

9.2 NUMERICAL SOLUTIONS TO ALGEBRAIC EQUATIONS

Chapter 2 was all about solving algebraic equations, using "Solve". Just as differential equations can be solved numerically with "NDSolve", algebraic equations can be solved numerically with "NSolve" and we'll review its use here.

Let's start with a simple example, finding the intersection between a line and an ellipse. This can be solved exactly using

```
eq1 = (x/a)^2 + (y/b)^2 == 1;
eq2 = c x + d y == 1;
exact = Solve[{eq1, eq2}, {x, y}];
```

where the expression returned in "exact" is indeed analytic, even though it looks very messy. If we put in numbers with

```
abcd = {a -> 1, b -> 2, c -> 3, d -> 4};
exact /. abcd
```

then we get answers for x and y in terms of fractions involving integers and square roots. Executing

```
N[exact /. abcd]
```

returns x and y as floating point numbers.

Now we could also make the numerical replacement in the equations and solve them instead with "NSolve", that is

```
NSolve[{eq1, eq2} /. abcd, {x, y}]
```

returns the same floating point numbers for x and y.

Of course, the real usefulness of "NSolve" is for algebraic equations that cannot be solved analytically. Consider trying to solve the equation

$$\sin(x) = \frac{1}{3}$$

for x. Using

```
Plot[{Sin[x], 1/3}, {x, 0, 2 Pi},
 PlotStyle -> {Solid, Dashed}]
```

we get

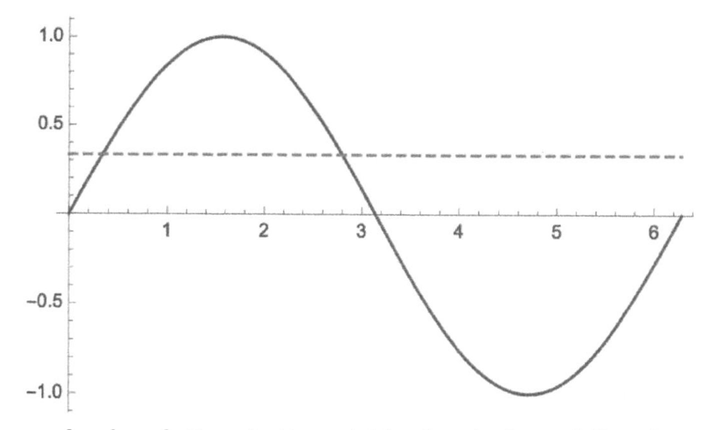

so there are clearly solutions in the neighborhood of $x = 1/3$ and $x = \pi - 1/3$, repeating with period 2π. Executing

```
Solve[Sin[x] == 1/3, x]
```

returns conditional expressions involving ArcSin functions. On the other hand

```
NSolve[Sin[x] == 1/3, x]
```

returns numerical values, also with conditional expressions. It is easy enough to extract the values relevant to the figure above, using

```
soln = NSolve[Sin[x] == 1/3, x];
soln /. C[1] -> 0
```

which returns

```
{{x -> 0.339837}, {x -> 2.80176}}
```

A different approach to solving this equation would be to find roots of the expression $\sin(x) - 1/3$. In MATHEMATICA this is done simply with the "FindRoot" function, for example

```
FindRoot[Sin[x] - 1/3, {x, 0.3}]
```

finds the root nearest $x = 0.3$, and returns

```
{x -> 0.339837}
```

9.3 WORKING WITH UNITS

In recent editions of MATHEMATICA, it is possible to work calculations directly with physical units attached. There is extensive documentation on this, and I don't necessarily recommend making heavy use of this until you are well versed in the basics, but it is good to be aware of the possibilities. For more information, see the Language Guide on *Units*.

Some MATHEMATICA functions actually return quantities with units, for example numbers retrieved from data bases. (See Section 9.4.) However, you can easily add units to numbers yourself. For example

```
len = Quantity[10.3, "ft"]
```

returns with "10.3 ft" which, although it looks like text, is actually a unit-containing quantity internal to MATHEMATICA. You can change the units on this quantity with something like

```
UnitConvert[len, "in"]
```

which returns "123.6 in". The default for unit conversion is always the SI system, also known as MKSA for the base units "meter-kilogram-second-Ampere". That is,

```
UnitConvert[len]
```

returns "3.13944 m".

You can use "Quantity" with no value, in which case it returns one in the specified units. That is

```
foot = Quantity["ft"]
```

returns "1 ft". If you specify a unit that is not a standard terminology, for example "gm" instead of "g" for *gram*, then MATHEMATICA will do its best to resolve your request and give you what it thinks you want. If you specify an incompatible unit, for example

```
UnitConvert[len, "kg"]
```

then you will get an error message.

The function "QuantityMagnitude" extracts the numerical value from the unit-containing quantity. This is one way you can switch a value from one system of units to another. For example

```
lenFeet = 10.3;
len = Quantity[10.3, "ft"];
lenInch = QuantityMagnitude[
   UnitConvert[len, "in"]];
lenInch/lenFeet
```

simply returns the number "12.". The function "QuantityUnit" returns the units associated with a unit-containing quantity.

Now here's something extremely useful for carrying out numerical calculations involving physical quantities. The "units" that MATHEMATICA knows are extensive, and include plenty of known values. For example, MATHEMATICA knows the speed of light, and you can use that as a unit. Therefore

```
cSI =
 QuantityMagnitude[
  UnitConvert[
   Quantity["SpeedOfLight"]]]
```

returns the value for the speed of light in km/sec, namely 299792458. A quick sampling of quantities I have found useful includes

"SpeedOfLight" (i.e. c)

"ElementaryCharge" (i.e. e)

"PlanckConstant" (i.e. h)

"ReducedPlanckConstant" (i.e. $\hbar$)

"BoltzmannConstant" (i.e. k)

"AtomicMassUnit" (i.e. u)

"ElectricConstant" (i.e. ϵ_0)

"MagneticConstant" (i.e. μ_0)

"GravitationalConstant" (i.e. G)

"ElectronMass"

"ProtonMass"

"BohrMagneton"

Remember that MATHEMATICA will try to figure out what you mean when you specify a unit, so if you need something, see if it is there. Most MATHEMATICA interfaces will also give you options to fill in the argument when you start typing, and that is helpful as well.

You can find more information in the MATHEMATICA Language Guide on *Scientific Data Analysis.*

9.4 ACCESS TO DATABASES

MATHEMATICA makes it possible for you to connect to external databases in any number of ways. We will just touch on these capabilities in this book. However, for detailed information, see the *Database Connectivity* Wolfram Language Guide.

In this section we will be particularly interested in the Wolfram Language Guide on *Physics & Chemistry: Data and Computation.* As it says in the Guide, this functionality "provides seamless access to the curated and continuously updated Wolfram Knowledgebase... which includes a wide range of types of data for physics and chemistry." For example, suppose you needed the lifetime and mass for the elementary particle known as the Λ hyperon. Then

```
ParticleData["Lambda", "Lifetime"];
LambdaLifetime = QuantityMagnitude[%]
ParticleData["Lambda", "Mass"];
LambdaMass = QuantityMagnitude[%]
```

returns in "LambdaLifetime" and "LambdaMass" the values 2.632×10^{-10} and 1115.683, respectively. These are the mean life of the Λ in seconds, and the mass of the Λ in MeV/c^2. (If you want the Λ mass in kg, then you'd need to use "UnitConvert" before extracting the magnitude.)

Some other specialized Physics & Chemistry quantities include "IsotopeData", "ElementData", and "ThermodynamicData". One *very* specialized quantity, useful for elementary particle detection or radiation shielding, is "StoppingPowerData". Another interesting one is "ChemicalData". For example

```
ChemicalData["Caffeine", "MoleculePlot"]
```

returns with the figure

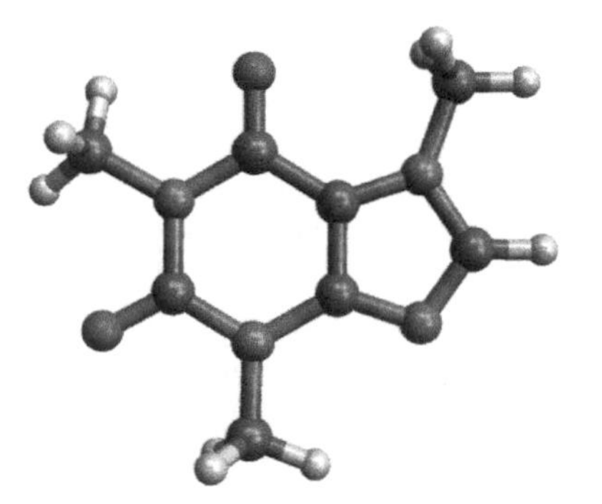

There are many databases available to you. If you ever need some value that you know is out there, use the MATHEMATICA Documentation Center to first check if it is at your fingertips, and it probably will be.

9.5 PHYSICS EXAMPLES

It is likely that you opened to this section first, so I'll take some extra time explaining these examples, and refer back to earlier material. The examples in this section are just for you to get your feet wet on numerical calculations. However, pick up just about any introductory physics textbook and go to the exercises at the end of just about any chapter, and you'll have plenty of other things to practice.

Example 9.1 *Calculate the nominal value of the acceleration g due to gravity at the Earth's surface, from the Earth's mass and nominal radius.*

First let's review this very simple physics. We write the force on a mass m at the Earth's surface as mg, but we know this is just the Newtonian gravitational force GmM/R^2 at the distance R from a mass M. So, we just solve this simple equation for g and put in the numbers.

Notebook 9.1 carries this through. I've included what you might consider some bad habits, using capital letters for some quantities, using assignments instead of replacements, and reassigning a replacement variable. However, the idea here is to just get to the answer quickly and not think of this as part of some larger problem.

Nevertheless, let's go through this solution line by line. As I've recommended, always start a new notebook with "Remove["Global'*]", at least while you're getting started, to wipe out any existing definitions.

As described in Chapter 2, we solve the equation $mg = gMm/R^2$ in the standard fashion. (Novices, note the argument to "Solve" is a logical construct with a double-equals sign. See Section 6.2 for an extended discussion of logical constructions.) Recall that the result returned in "sol" is a nested list, but this time of just one variable and with just one answer.

After solving the equation, I extract just the replacement statement for "g" using the shorthand for "Part" (Recall Section 2.2) and use it to redefine "g" as the expression $g = GM/R^2$. (If you're not familiar with all this remove the semicolons at the ends of the two lines discussed here, and let MATHEMATICA print out the results for you.)

Earlier in this chapter, in Section 9.4, I talked about how to use the function "Quantity" to extract known values of physical quantities. In this case, we need the values of G, M, and R, and I follow the same prescription to get them in SI units. (As you type in ""EarthMeanRadius"", you'll notice that MATHEMATICA gives you options, and one of them is ""EarthEquitorialRadius"", but we choose the mean radius instead.) We use straightforward assignment statements to set the values of these three quantities.

Finally, we simply ask for the numerical value of g with standard precision. The result, in SI units as 9.8 m/sec^2, is of course what we all know from our first course in physics.

Example 9.2 *A nuclear reactor gets its energy mainly from the fission of the ^{235}U nucleus after it absorbs a very low energy neutron. A typical reaction is*

$$n +^{235} \mathrm{U} \rightarrow^{92} \mathrm{Kr} +^{144} \mathrm{Ba}$$

Calculate the number of fissions per second in a 500 *MW reactor. How long does it take to burn up 10 kg of ^{235}U?*

This is a very simple exercise, realizing that the energy of a nuclear reactor comes from the excess mass converted to energy, according to $E = mc^2$.

Notebook 9.2 shows the calculation. After retrieving the SI value for c, the notebook gets the atomic masses (again in their SI values) for the masses of ^{236}U,^{92}Kr, and ^{144}Ba, and converts them to energy with a multiplication by c^2. The mass of ^{236}U is held in a separate object "m236" because we'll need it later.

NOTEBOOK 9.1 A very straightforward solution to Example 9.1.

```
In[1]:= Remove["Global`*"]
```

Gravitational acceleration on Earth

Set up and solve the basic equation

```
In[2]:= sol = Solve[m g == G M m / R^2, g];
        g = g /. sol[[1]];
```

Put in the numbers

```
In[4]:= Quantity["GravitationalConstant"];
        UnitConvert[%];
        G = QuantityMagnitude[%];

In[7]:= Quantity["EarthMass"];
        UnitConvert[%];
        M = QuantityMagnitude[%];

In[10]:= Quantity["EarthMeanRadius"];
         UnitConvert[%];
         R = QuantityMagnitude[%];

In[13]:= N[g]

Out[13]= 9.8196
```

NOTEBOOK 9.2 The calculation called for in Example 9.2.

```
In[1]:= Remove["Global`*"]
```

Energy from 235U Fission

Get the isotope data

```
In[2]:= c = QuantityMagnitude[
          UnitConvert[Quantity["SpeedOfLight"]]];
        m236 = QuantityMagnitude[UnitConvert[
           IsotopeData["Uranium236", "AtomicMass"]]];
        E236 = m236 * c^2;
        E092 = QuantityMagnitude[UnitConvert[
            IsotopeData["Krypton92", "AtomicMass"]]] * c^2;
        E144 = QuantityMagnitude[UnitConvert[
            IsotopeData["Barium144", "AtomicMass"]]] * c^2;
```

Calculate the energy released in one fission

```
In[7]:= Efission = E236 - E092 - E144;

In[8]:= e = QuantityMagnitude[
          UnitConvert[Quantity["ElectronCharge"]]];
        (Efission / e) 10^(-6)

Out[9]= 183.0
```

Calculate the fission rate and burnup time

```
In[10]:= rate = 500 × 10^6 / Efission

Out[10]= 1.705 × 10^19

In[11]:= time = (10 / m236) / rate;
         time / (24 × 3600)

Out[12]= 17.32
```

The energy in one fission is then calculated. We take a little detour to convert this number into MeV "the old fashioned way", by dividing by the electron charge. This is a good check. It is generally known that fission of a heavy element yields around 200 MeV in energy.

Finally, the fission energy is divided into 500×10^6 Watts to get a rate of 1.7×10^{19} fissions per second. Each fission uses up one ^{236}U nucleus, so we divide 10 kg by the mass to find the number of ^{236}U nuclei. Dividing the rate into this number gives 17.3 days for a burnup time.

Example 9.3 *The mass of the Sun is much larger than any of the planets. Assuming circular orbits, estimate the solar mass from data for each planet. Plot the ratio of this estimate to the accepted value of the solar mass, as a function of distance of the planet to the Sun.*

For a planet of mass m orbiting the Sun with mass M, we apply Newton's Second Law with Newtonian gravity to obtain

$$G\frac{Mm}{r^2} = m\frac{v^2}{r} \qquad \text{so} \qquad M = \frac{rv^2}{G}$$

where r is the radius of the (circular) orbit, and $v = 2\pi r/T$ is the speed of the planet with orbital period T. So, the calculation is simple, and most of the work is figuring out where to get the data.

Notebook 9.3 shows how to do this. The key database is accessed using the function "PlanetData". The first step is to just create a list with the planet names. I then execute "PlanetData["Properties"]" which provides a list of all the different data options. (The output is suppressed to save space, but I urge you to try this for yourself.) I use "AverageOrbitDistance" to get the orbit "radius", and "OrbitPeriod" for the orbital period. These quantities are not necessarily in SI units - the orbit distances in fact are in Astronomical Units - so we convert the quantities to their SI form before extracting the numbers.

Newton's gravitational constant G is a "unit" so it is easy to extract its value in SI units. We then calculate the speeds, and use this to calculate the solar mass for each planet.

I suppressed the output in the notebook, but I'm including the plot here:

NOTEBOOK 9.3 A very straightforward solution to Example 9.3.

```
In[1]:= Remove["Global`*"]
```

Mass of the Sun from planet data

Get the planet data

```
In[2]:= planets = {"Mercury", "Venus", "Earth", "Mars",
          "Jupiter", "Saturn", "Uranus", "Neptune"};

In[3]:= PlanetData["Properties"]

In[4]:= planetDistances = QuantityMagnitude[
          UnitConvert[
           PlanetData[planets, "AverageOrbitDistance"]]];

In[5]:= planetPeriods = QuantityMagnitude[
          UnitConvert[
           PlanetData[planets, "OrbitPeriod"]]];
```

Calculate solar mass from data

```
In[6]:= G = QuantityMagnitude[
          UnitConvert[Quantity["GravitationalConstant"]]];
        v = 2 Pi planetDistances / planetPeriods;
        solarMasses = planetDistances v^2 / G;
```

Plot relative to accepted value of solar mass

```
In[9]:= solarMass = QuantityMagnitude[
          UnitConvert[Quantity["SolarMass"]]];

In[10]:= points = Transpose[
           {Log10[planetDistances], solarMasses / solarMass}];
         ListPlot[points, PlotRange → {{10.5, 13.0}, {0.99, 1.07}},
          AxesLabel → {"Log(Distance/m)", "Calculated/True"},
          PlotMarkers → {Automatic, Large}]
```

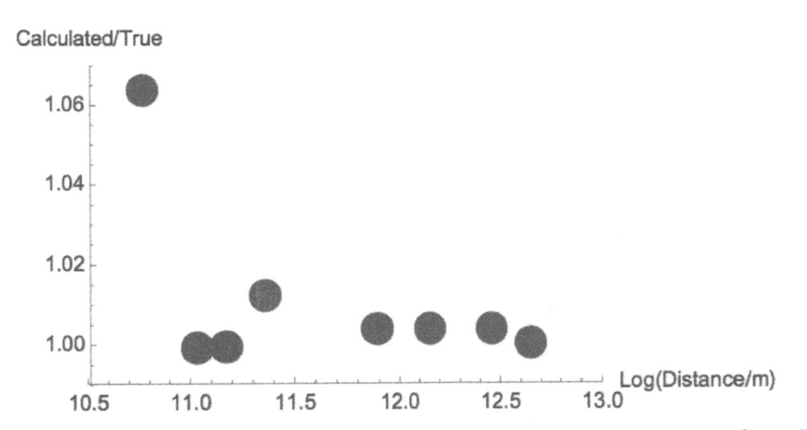

Making the plot is straightforward, nothing we haven't used before. The solar mass is also a "unit", so we extract it, and use it to form the data set before plotting. Note that I am plotting the calculated mass versus the *log* of the distance, only because it spreads the points out more nicely. You'll see that Mercury gives a value that is more than 6% high, but the other planets give values within a couple of percent. Notice that all the points are a little higher than the accepted value. A true calculation needs to take into account that the orbits are actually elliptical, and that the reduced mass effect can be significant.

9.6 CHAPTER SUMMARY

- In addition to its use for symbolic manipulation, MATHEMATICA also has facilities for working with numerical data with high precision.
- One facility is numerical solution of algebraic equations, including root-finding functionality. (We have already seen how to numerically solve differential equations.)
- It is possibly to identify "quantities" that are attached to units. Many physical quantities can be identified by name. A unit conversion function is also provided.
- MATHEMATICA gives you access to a large number of physical and chemical databases. The interface is simple to use.

EXERCISES

9.1 *How long does it take for a rock to fall from 2m height on the Moon? You'll first want to follow the example in Notebook 9.1 to find the local gravitational acceleration on the Moon. The function "*PlanetaryMoonData*" will get you the mass and radius, but be careful of the default units. (Of course, you can use "*UnitConvert*" to put everything in the SI, or some other, system of units.) Simple one dimensional kinematics will help you determine the time* *it takes the rock to fall. You can also use this opportunity to get some practice with equation solving and other simple operations in* MATHEMATICA.

9.2 *Two electrically neutral 100 gram iron balls are separated by a distance of one foot. How many electrons have to move from one ball to another so that the force between the balls is ten pounds? This number is what fraction of the total number of electrons on one of the balls?*

Try to use MATHEMATICA *to look up the various quantities and conversion factors. The value of* ϵ_0 *used in Coulomb's Law is called* "VacuumPermittivity" *in* MATHEMATICA. *You may need to review some freshman chemistry to find the number of iron atoms. Use the function* "ElementData" *to look up the atomic number and atomic weight of iron. Note that one of the "units" for quantities is Avogadro's number and the SI version is just the number itself.*

You might be surprised to see how small a fraction of electrons need to be transferred in order to create such a large force.

9.3 *A certain high power pulsed laser delivers 2.4 mJ of energy in a 50 fs time window over a 0.4* mm^2 *area. Find the peak electric field contained in this laser pulse, and compare it to the electric field in a hydrogen atom at one Bohr radius from the proton.*

9.4 *A damped harmonic oscillator with mass* $m = 0.05$ *kg, spring constant* $k = 1.2$ *N/m, and damping coefficient* $b = 0.2$ *N·sec/m, starts from rest at a point 0.75 m from equilibrium. Find the first two times at which the mass crosses the equilibrium point.*

9.5 *What is the wavelength of light emitted in the* $n = 3 \rightarrow n = 2$ *transition in singly ionized helium?*

9.6 *The energy density* $\rho(\lambda)$ *of "blackbody" cavity radiation, in energy per unit volume per unit wavelength* λ*, is given by*

$$\rho(\lambda) = \frac{8\pi hc}{\lambda^5} \frac{1}{e^{hc/\lambda kT} - 1}$$

where h *is Planck's constant,* k *is Boltzmann's constant,* c *is the speed of light, and* T *is the cavity temperature. Find an expression for the wavelength at which the density is a maximum. (You will need to form a dimensionless variable in order to numerically solve an algebraic equation.) Also find an expression for the "radiance", which is* $c/4$ *times the integral of the energy density over all wavelengths. Put in values for* h*,* c*, and* k*, and compare to the answers found in any Modern Physics textbook.*

CHAPTER 10

Random Numbers

CONTENTS

Physicists often find it useful to generate sets of random numbers to simulate experimental data or to carry out otherwise difficult integrals. This chapter shows you the basics of random number generation.[1]

You should realize that computer generated "random" numbers are not actually random. They are the products of algorithms which are the result of lots of research on how best to generate series of integers or real numbers that are effectively random. However, for most of the problems you'll ever want to solve, this won't concern you.

So called Monte Carlo techniques are very common in physics. This is essentially a method of doing complicated, multidimensional integrals, although physicists don't usually cast it in that form. Rather, random numbers are generated in a way that samples the space in which certain kinds of "events" can happen. In this way, you can simulate real, physical situations which are dominated by some kind of random process. Examples might include the effect of cosmic rays on some detector geometry, or the process by which atoms adsorb onto the surface of some kind of substrate.

Monte Carlo simulations will generally require that the random numbers be generated according to some probability distribution. That is, you might want to generate random numbers between two limits, but not with the same probability at each place in the interval. Cosmic rays, for example, mostly come downward, with very few at large zenith angles. If you are simulating a cosmic ray detector, then, you want to have a realistic distribution of cosmic ray angles.

All these techniques are discussed in this chapter. The Physics Example

[1] The Wolfram Language Overview on *Random Number Generation* provides a detailed discussion of the concepts in this chapter.

makes use of a "random walk", which might be adapted to any number of applications.

10.1 GENERATING RANDOM NUMBERS

Algorithms used to generate (seemingly) random numbers are rather complicated, but we won't concern ourselves with that here. To be sure, computers generate so-called *pseudo*-random numbers, in patterns that can be reproduced if necessary, but which nevertheless demonstrate a sufficiently high degree of independence.

Here we focus on the two functions "RandomReal", which generates a pseudo random real number, and "RandomInteger", for generating integers. They each can be used with different arguments that will return a single number or a list of numbers. For example

```
RandomReal[{1, 5}]
```

returns a single, pseudorandom real number r where $1.0 \leq r \leq 5.0$, while

```
RandomInteger[{1, 5}, 4]
```

returns a list of four pseudorandom integers k where $1 \leq k \leq 5$. In two successive trials of this command, MATHEMATICA returned for me

```
{4, 1, 3, 4}
```

and then

```
{5, 4, 3, 5}
```

See the Language Tutorial *Random Number Generation* for information on setting the "seed" of the pseudorandom number generator so that you can reproduce the sequence of numbers. You will also find more variations on these two functions, and also how to generate pseudorandom complex numbers and prime numbers.

We will see later in this chapter how to generate random numbers according to given probability distributions.

10.2 MONTE CARLO TECHNIQUES

Monte Carlo is a small region on the Mediterranean Sea which is known worldwide as a center for extravagant gambling. Somehow, the name was borrowed to describe techniques for using random number generation to simulate physical situations.

At its essence, the Monte Carlo method is a way to do an integral. Its usefulness is mainly for doing very complicated, multidimensional integrals, but it is easy to understand the principle using a one dimensional integration. You know from your first course in calculus that

$$\int_a^b f(x)dx = \text{Area under the curve}$$

That is, the integral is the area of the figure bounded by the x-axis, the curve $f(x)$, and the vertical lines at $x = a$ and $x = b$.

This integral can be estimated using randomly generated (x, y) pairs of data. First generate x values where $a \leq x \leq b$. Then generate y values between the minimum (f_{min}) and maximum (f_{max}) values of $f(x)$ (in the interval $a \leq x \leq b$). These random numbers populate an area $(b-a)(f_{\text{max}} - f_{\text{min}})$ and if you count up the number of pairs that fall within the curve, that fraction of the total number generated, times the rectangular area, gives you the area under the curve.

Let's illustrate this with a simple example. The area of a quarter-circle with unit radius is $\pi/4$. Therefore,

$$\pi = 4 \int_0^1 (1 - x^2)^{1/2} dx$$

and we can estimate the value of π using a Monte Carlo technique. First, we generate points that randomly populate the square bounded by $0 \leq x \leq 1$ and $0 \leq y \leq 1$. Then, we count the number of points that fall under the curve. Finally, carry out some simple arithmetic to get our approximation to π.

The following cell generates the random numbers and produces a plot that illustrates the approach:

```
nGen = 5 10^3;
points = RandomReal[{0, 1}, {nGen, 2}];
Show[
 ListPlot[points,
  PlotRange -> {{0, 1}, {0, 1}}],
 Plot[Sqrt[1 - x^2], {x, 0, 1},
  PlotStyle -> Thick],
 AspectRatio -> 1]
```

For the plot, I make sure that graph looks like an appropriate square and the line looks like a circle by forcing the plot limits and setting the aspect ratio to unity. The output is

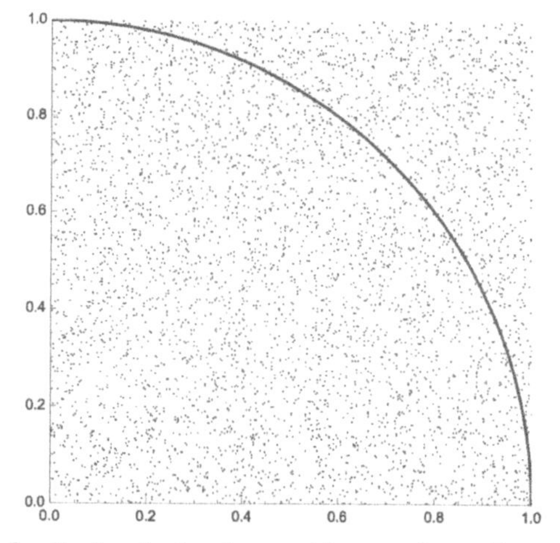

Note that I make it simple to change the number of generated points from 5000 to something else. The larger the number of samples, the more accurate I expect to be my approximation, but the longer it will take to run the program. Also, I use an alternate form of "RandomReal" to create the set of points, one where you can fill an array of arbitrary dimensions with random numbers.

Now we need to count the points under the curve. Recall the discussion in Section 7.2 on using "Select" to apply "cuts" to data. We can count the points using the "area under the curve" with the command

```
nPi = Length[Select[points, #[[2]] < Sqrt[1 - #[[1]]^2] &]]
```

We can also recognize that we are counting the points inside the circle, and execute

```
nPi = Length[Select[points, Sqrt[#[[1]]^2 + #[[2]]^2] < 1 &]]
```

A slicker way to count the number of points inside the circle is to treat each point as a two-dimensional vector and use

```
nPi = Length[Select[points, Dot[#, #] < 1 &]]
```

All three of these, of course, return the same value for "nPi", which is "3931" for the time I executed my notebook, that is, for the random numbers I generated at the time.

All that remains is to calculate π. Executing

```
N[4*nPi/nGen]
100 (1 - %/N[Pi])
```

returns "3.1448" and tells me that this is 0.10% too low.

10.3 PROBABILITY DISTRIBUTIONS

If we generate a total of N random numbers with values x between x_{lo} and x_{hi}, then the probability of finding a value between x and $x+\Delta x$ is $\Delta x/\left(x_{\text{hi}} - x_{\text{lo}}\right)$.

In other words, we should find approximately $N\Delta x/\left(x_{\rm hi}-x_{\rm lo}\right)\equiv\Delta N$ values in this range. Formally, we might write $f(x)=1/\left(x_{\rm hi}-x_{\rm lo}\right)$ (a constant) and call it a *probability distribution function.* We would then write

$$\Delta N \approx Nf(x)\Delta x \qquad \text{or} \qquad \frac{\Delta N}{\Delta x} \approx Nf(x) \tag{10.1}$$

where the equality holds if Δx is small enough, and

$$\int_{x_{\rm lo}}^{x_{\rm hi}} f(x)dx = 1 \tag{10.2}$$

Of course, this can be generalized to any $f(x)$, so long as it meets the normalization condition (10.2).

You can think of a histogram of values $\Delta N/\Delta x$, binned in values of x, as an approximation to $f(x)$. In fact, you can convince yourself that a very large sample can be used to trace out the shape of $f(x)$. Physicists will frequently carry out a fit to the shape of a histogram, in order to get an idea of the probability distribution $f(x)$.

How do we generate random numbers that follow an arbitrary probability distribution $f(x)$? The key is to use the inverse of the antiderivative of $f(x)$. I will leave a more formal discussion to other books, but here is the basic idea.

First, let's take $x_{\rm lo}\to-\infty$ and $x_{\rm hi}\to+\infty$. This is perfectly general, since we can always set $f(x)=0$ in particular regions. Now define

$$F(x) = \int_{-\infty}^{x} f(t)dt \tag{10.3}$$

The function $f(x)$ is positive everywhere, and $F(-\infty)=0$ and $F(+\infty)=1$, so $0\le F(x)\le 1$. If I set $F(x)=r$, where r is a random number between zero and one, and solve for x using the inverse function, that is $x=F^{-1}(r)$, then x will be distributed according to $f(x)$.

I'm asserting this to be true without proving it, but you can get a feeling that it is a correct prescription. In regions where $f(x)$ is close to zero, $F(x)$ is not changing, so no value of r will return a value of x. On the other hand, near a maximum of $f(x)$, $F(x)$ has a very high slope, and there are many values of r that will return x. I encourage you to learn more about this technique, but we will leave this discussion at this point.

Let's illustrate this with the Gaussian or Normal Distribution, namely

$$f(x;\mu,\sigma) = \frac{1}{\sigma\sqrt{2\pi}}e^{-(x-\mu)^2/2\sigma^2} \tag{10.4}$$

The factor in front of the integral is necessary to meet the normalization condition (10.2). This distribution is commonly used in statistical analysis. It is a symmetric distribution about the mean at $x=\mu$ with standard deviation

σ. Following Equation 10.3 we define the function

$$\begin{aligned} F(x;\mu,\sigma) &= \int_{-\infty}^{x} f(y;\mu,\sigma)dy = \frac{1}{\sqrt{\pi}}\int_{-\infty}^{(x-\mu)/\sigma\sqrt{2}} e^{-t^2}dt \\ &= \frac{1}{\sqrt{\pi}}\left[\int_{-\infty}^{0} e^{-t^2}dt + \int_{0}^{(x-\mu)/\sigma\sqrt{2}} e^{-t^2}dt\right] \\ &= \frac{1}{2}\left[1+\text{erf}\left(\frac{x-\mu}{\sigma\sqrt{2}}\right)\right] \end{aligned} \tag{10.5}$$

where $\text{erf}(x) \equiv (2/\sqrt{\pi})\int_0^x e^{-t^2}dt$ is called the "error function". So, in order to generate a random number according to the distribution in Equation 10.4, we set $F(x;\mu,\sigma)$ equal to a random number between zero and one, and solve for x. Equivalently, we set $\text{erf}[(x-\mu)/\sigma\sqrt{2}]$ to a random number between -1 and 1, and solve for x.

MATHEMATICA provides us with "Erf" to calculate the error function, and "InverseErf" to calculate its inverse. It is therefore simple to come up with a list of random numbers distributed according to a Gaussian. For example

```
mu = 1.5;
sigma = 2.5;
rangs = mu + sigma Sqrt[2] InverseErf[RandomReal[{-1, 1}, 1000]];
```

generates a set of 1000 numbers "rangs" with a mean 1.5 and standard deviation 2.5. You should execute these statements, and then use "Mean" and "StandardDeviation" to confirm that you get reasonable answers, perhaps also histogram them and compare to Equation 10.4. To write these numbers to the data file used in Section 7.3, execute

```
Export["rangs.dat", rangs]
```

Just be aware of the directory in which MATHEMATICA deposits the file.

Now that I've spent some effort describing the nuts and bolts of generating random numbers according to a (one dimensional) probability distribution, I can tell you how to do this much more simply, thanks to the extensive capabilities of MATHEMATICA when it comes to statistical analysis.

Many standard Probability Distribution Functions (PDFs) are built into MATHEMATICA. You can learn about them from the tutorials on *Continuous Distributions* and *Discrete Distributions.* I will show you one example of each, and leave it to you to learn about the others.

For an example of the continuous distribution, let's just use the Gaussian (or Normal) distribution described earlier in this section. To generate a list of 1000 random numbers with a mean 1.5 and standard deviation 2.5, that is, with the same parameters as the example earlier in this section, execute

```
mu = 1.5;
sigma = 2.5;
npoints = 1000;
ranGauss = RandomVariate[
   NormalDistribution[mu, sigma], npoints];
```

Two new functions are included here. The function "NormalDistribution[μ,σ]" is a special kind of MATHEMATICA function that "represents" a Gaussian distribution with mean μ and standard deviation σ. It is generally used as an input to other functions. Indeed, here it is used as input to the second new function "RandomVariate", which generates a random number, or a list of random numbers, according to the specified probability distribution function.

From one run of the above cell, executing

```
Mean[ranGauss]
StandardDeviation[ranGauss]
```

returns "1.6016" and "2.52001", consistent with the parameters I used to generate the random numbers in the first place.

A third new function that takes a probability distribution as input is "PDF", which returns a pure function based on the mathematical expression represented by the input. We can put this together with what we already know, to get a graphical test of our random number generation.

Let's make a histogram of the random numbers, and overlay on top a plot of the expected distribution. Remember that the distribution is normalized to unity, so we'll need to multiply it by the appropriate numbers to make the comparison. We can accomplish this by executing the cell

```
pdf = PDF[NormalDistribution[mu, sigma]][x];
binwidth = 0.2;
Show[
 Histogram[ranGauss, {binwidth}],
 Plot[npoints pdf binwidth, {x, -5, 10}]]
```

which produces the following plot:

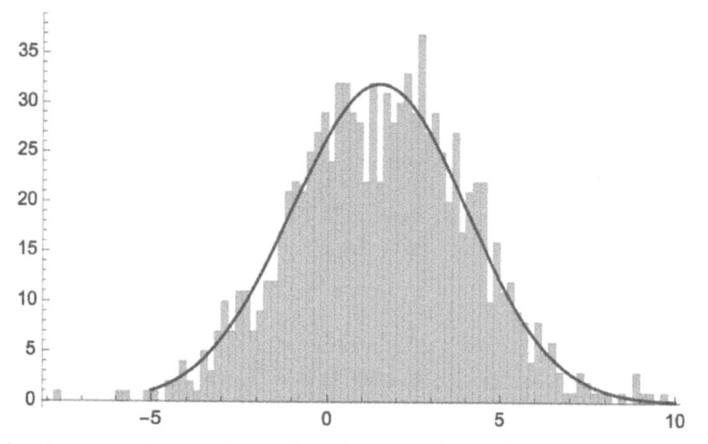

The overlay is created by using "Show" to combine "Histogram" and "Plot", in that order, so that the plot curve sits "on top" of the histogram. We specify the bin width for the histogram as 0.2. (Recall Section 7.4 on creating histograms.) We use "PDF" to generate an expression of the Gaussian distribution as a function of x, and then plot this expression, after normalizing it by the number of points and bin width.

The comparison between the plot and the histogram appears satisfactory, but there is a caveat. We have used the default output for "Histogram" which uses the left edge of the bin for the horizontal axis, and we interpret this above as the value for x in the distribution. We should actually use the center of the bin for x, but in this case, the bins are narrow enough so that this doesn't matter much.

I suggest an exercise for you. Make this correction, and plot the curve as a function of the bin centers. See how much of a difference this makes if you have wider bins.

Next we'll do an example of a discrete distribution, that is, a probability distribution which can only return discrete values. There are two good introductory examples. One is the the Binomial Distribution, which gives the number k of successes, given the probability $p < 1$ of a single success and n trials. The second is the Poisson Distribution, which is an approximation to the Binomial Distribution for $p \ll 1$ and $n \to \infty$ with $\mu = np$ fixed. It should be clear that μ is the mean number of successes.

Let's choose the Poisson Distribution to demonstrate "RandomVariate" and "PDF" with discrete distributions. The Poisson Distribution is given by

$$P_\mu(k) = \frac{\mu^k}{k!} e^{-\mu} \tag{10.6}$$

and you should confirm for yourself that the normalization condition (for a discrete distribution), namely

$$\sum_{k=0}^{\infty} P_\mu(k) = 1$$

is satisfied. You can also prove that the standard deviation of $P_\mu(k)$ is $\sqrt{\mu}$. In fact, it is this simple property that makes the Poisson distribution so useful for analysis of physical data. Executing the cell

```
mu = 1.8;
npoints = 250;
ranPoiss = RandomVariate[PoissonDistribution[mu], npoints];
```

generates 250 random integers distributed according to a Poisson distribution with mean 1.8. The commands

```
N[Mean[ranPoiss]]
N[Variance[ranPoiss]]
```

returned "1.688" and "1.67737" for my set of random integers. (The *variance* is the square of the standard deviation, so should be approximately the same as the mean for a Poisson distribution of random integers.) I've used "N" to get real values for the mean and variance, because with these integer values, "Mean" and "Variance" by themselves return integer fractions.

Now let's make a plot to compare the true Poisson distribution with our random numbers. It is straightforward to get an expression for the Poisson distribution using

```
pdf = PDF[PoissonDistribution[mu]][x];
```

We want to compare this to a plot of the actual frequency distribution of the points we generated, but a "histogram" plot isn't quite right. With a discrete distribution, we are not "binning" continuous data, but rather just want points to indicate the number of zeroes, ones, twos, and so on.

We can use "HistogramList", also discussed in Section 7.4, to come up with these points and feed them to "ListPlot". Recall that "HistogramList" returns two lists, one of which is the bin edges and the other the frequencies. There is one more element in the first list than the second, so we need something like

```
bins = HistogramList[ranPoiss][[1]];
vals = HistogramList[ranPoiss][[2]];
xbin = Part[bins, 1 ;; Length[bins] - 1];
```

to get equal length lists for the bins and the frequency values. Then execute

```
Show[
 ListPlot[Transpose[{xbin, vals}],
  PlotStyle -> PointSize[Large]],
 Plot[npoints pdf, {x, 0, 8}]]
```

to plot the points and the distribution together, where I've opted to make the point size "large" to more easily set it off from the curve. The output is

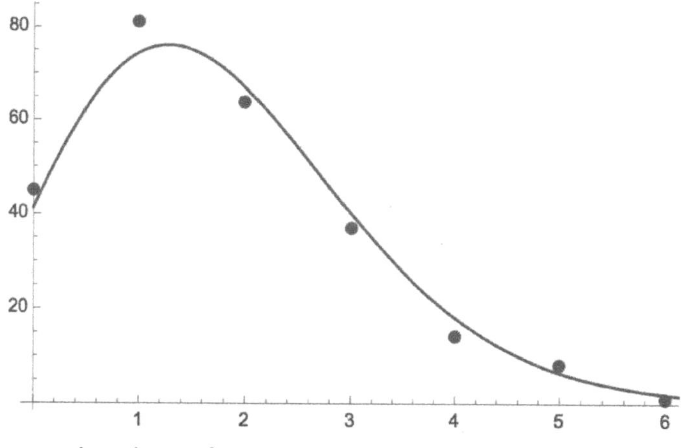

and the comparison is good.

10.4 PHYSICS EXAMPLE

Example 10.1 *A large number of ants are located at one point. Each takes a 1 mm long step in a random direction. Then, from their new position, they each take another 1 mm step in a random direction. The process repeats. How many steps does it take for half of the ants to leave a circle with radius 1 cm?*

This is a classic example of a "Random Walk" problem. We'll go through a solution that does more than just arrives at the answer, but also provides some tests along the way. You might notice, by the way, that the question asked is ambiguous, since an ant can leave the circle but then wander back in.

Notebook 10.1 provides a solution. We'll go through it step by step, but remember that since we are generating random numbers, you won't get the same numerical results as I get, but they should nevertheless be consistent.

First we set up the problem. We choose to follow 1000 ants, each for 200 steps. Each step is radially outward by one unit, $r = 1$ mm, as specified by the problem. The radius at which we'll test for "escape" is ten times larger.

Next we generate the random steps. The approach is to move each ant a horizontal distance $\Delta x = r\cos\theta$ and a vertical distance $\Delta y = r\sin\theta$, where θ is a two-dimensional array of numbers between 0 and 2π. One dimension of the array counts ants and the other dimension counts steps.

You may wonder why "theta" is dimensioned "(nSteps, nAnts)" instead of the other way around. The answer is that it doesn't matter. For the upcoming analysis, we just need to be aware and transpose the list if we are looking at a particular ant or a particular path. It is a good idea to remove the semicolons (and choose smaller numbers of ants and steps) so that you can see clearly how the list is organized. This will help you formulate the commands to follow.

In order to add up the steps, we use the function "Accumulate", which produces a list that is the cumulative running sum of the input list. That is,

NOTEBOOK 10.1 Solution to the random walk of Example 10.1

```
In[1]:= Remove["Global`*"]
```

Ants executing a random walk

```
In[2]:= nAnts = 1000;
        nSteps = 200;
        stepSize = 1;
        radius = 10;
```

Calculate the walk

```
In[6]:= theta = RandomReal[{0, 2 Pi}, {nSteps, nAnts}];
        xSteps = stepSize Cos[theta];
        ySteps = stepSize Sin[theta];
        xWalk = Accumulate[xSteps];
        yWalk = Accumulate[ySteps];
```

Plot the walk of the first ant

```
In[11]:= xVals = Flatten[{0, Transpose[xWalk][[1]]}];
         yVals = Flatten[{0, Transpose[yWalk][[1]]}];

In[13]:= antPath = Transpose[{xVals, yVals}];
         Show[
          ListPlot[antPath, Joined → True, PlotMarkers → {Automatic, Small},
           PlotRange → {{-12, 12}, {-12, 12}}, AspectRatio → 1],
          ParametricPlot[{radius Cos[u], radius Sin[u]}, {u, 0, 2 Pi},
           PlotStyle → Red]]
```

Analyze the collection of ants

```
In[15]:= GraphicsRow[{
           ListPlot[Transpose[{xWalk[[nSteps / 4]], yWalk[[nSteps / 4]]}],
            PlotRange → {{-50, 50}, {-50, 50}}, AspectRatio → 1],
           ListPlot[Transpose[{xWalk[[nSteps / 2]], yWalk[[nSteps / 2]]}],
            PlotRange → {{-50, 50}, {-50, 50}}, AspectRatio → 1],
           ListPlot[Transpose[{xWalk[[nSteps]], yWalk[[nSteps]]}],
            PlotRange → {{-50, 50}, {-50, 50}}, AspectRatio → 1]},
          ImageSize -> Full]

In[16]:= rWalk = Sqrt[xWalk^2 + yWalk^2];
         histLimit = 3 radius
         GraphicsRow[{
           Histogram[rWalk[[nSteps / 4]], {0, histLimit, 0.5}],
           Histogram[rWalk[[nSteps / 2]], {0, histLimit, 0.5}],
           Histogram[rWalk[[nSteps]], {0, histLimit, 0.5}]},
          ImageSize -> Full]

In[19]:= ListPlot[Mean[Transpose[rWalk]]]
         ListPlot[StandardDeviation[Transpose[rWalk]] / Mean[Transpose[rWalk]]]
         Length[Select[Map[Max, Transpose[rWalk]], # < radius &]]
         Length[Select[rWalk[[nSteps]], # < radius &]]
```

```
Accumulate[{a, b, c, d}]
```

returns

```
{a, a + b, a + b + c, a + b + c + d}
```

When acting on a nested list, Accumulate treats the inner lists as independent objects and forms the cumulative sum, so that

```
Accumulate[{{1, 2, 3, 4}, {a, b, c, d}}]
```

returns

```
{{1, 2, 3, 4}, {1 + a, 2 + b, 3 + c, 4 + d}}
```

This means that the objects "xWalk" and "yWalk" are two dimensional arrays that represent the step-by-step positions of each of the ants.

Next we study the result. It is instructive to show the actual path of an arbitrary ant. We make two new lists "xVals" and "yVals" that mark the path of an ant. For this, we need to transpose the lists and use "Part", in shorthand, to pick out the first ant. (The first ant is as arbitrary as any other.) We add "0" to the head of each list so that the ant starts at the origin, using "Flatten" to create a single one-dimensional list.

The next cell plots the ant's path. Remember that every time you execute the notebook, the random numbers are different. For one execution of the notebook, I get the following:

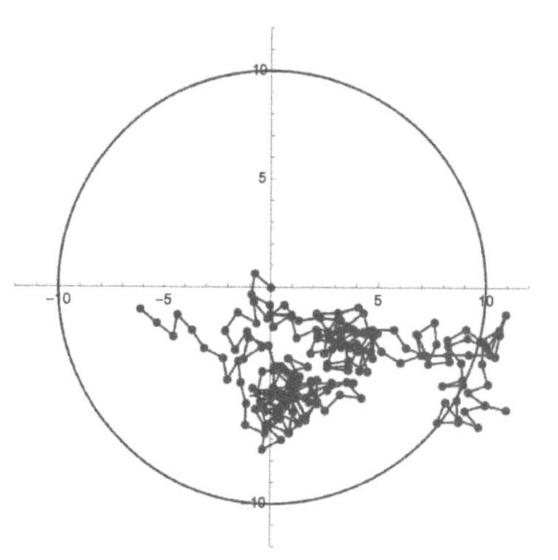

This particular ant wanders south, hanging out in the southeast quadrant, leaves the marked circle for a while, then reenters and ends up west of the origin, just below the x-axis near $x = -6$.

Let's look carefully at the cell that makes this plot. We'll plot the individual positions after each step, so we make a two-dimensional list "antPath" from the x and y positions. We plot these positions using "ListPlot", joined by line segments, but still including the default plot markers for the points. A square plot range is specified, along with a square aspect ratio, so the plot appears how it would look on a table top.

We use "Show" to combine the "ListPLot" of the path, along with the circle (in "Red") that is specified in the problem statement. The circle is drawn with "ParametricPlot", although later we will learn more succinct ways to draw circles and other fundamental shapes.

Now on to analyzing the data we've generated. A good first step, which is a useful check that we have generated the paths correctly[2], is to plot the positions of all the ants after different numbers of steps, just to see if the pattern increases in size as you'd expect.

The next cell makes these plots, horizontally next to each other using the function "GraphicsRow", with an option for a "Full" image size. In addition to the ending point, steps are chosen at 1/4 and 1/2 of the total number. The plots are made all with the same limits and, again, a square aspect ration. The result is

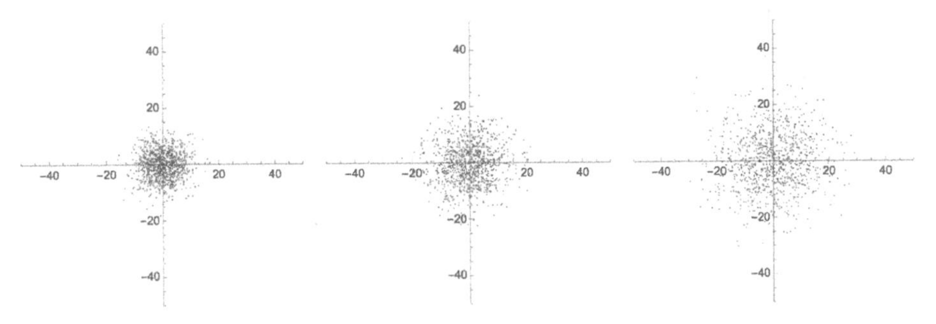

This looks about right. The endpoints are symmetric about the origin, and they seem to spread out monotonically as the number of steps increases. In fact, the pattern looks to be the same at each number of steps, except that the scale increases. This illustrates the notion of "scale invariance".

The problem statement asks "*How many steps does it take for half of the ants to leave a circle with radius 1 cm?*", but what does that mean? As we've seen, an ant can leave the circle and then come back in. Do we want to know how many ants *ever* leave the circle, or how many end up outside the circle? Or, is it a statistical question, that is, when does the mean distance from the center exceed the specified radius?

Instead of deciding which of these questions to answer, let's investigate the general pattern. (You should take this notebook and modify it so you can answer the questions in your own way.) The next cell first defines "rWalk" as the distances from the origin for each ant at each step. It then makes another three-across figure, plotting histograms of these distances for all the ants at the same three step counts as the figure above. The histograms are made with the same bin specification so that they can be compared more easily, that is

[2]When I first wrote this notebook, I made an error in the path generation, which I didn't catch until I made this plot.

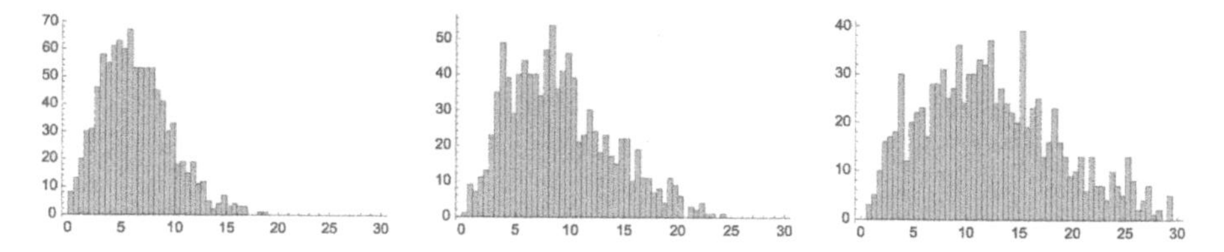

From the histograms, it is clear that somewhere between one-half and the full number of steps, the mean number of ants outside the circle crosses 50%. The final cell in Notebook 10.1 gets more quantitative about this. First it creates two more plots. One is just the mean of all the ants' radial positions as a function of the step. The other is the ratio of the standard deviation to the mean, which should provide some information about the shape of the distributions. The resulting two plots are

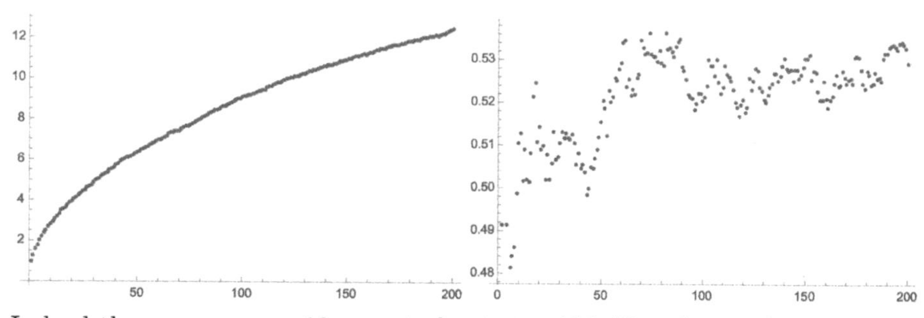

Indeed the mean crosses 10 mm at about step 120. The plot on the right tells us something very interesting about the shape, regarding "scale invariance", namely that the shape really doesn't change much, if at all. Note the vertical scale. The standard deviation as a fraction of the mean remains very nearly constant as the number of steps increases.

Finally, the last two commands in the last cell tell us the number of ants that stepped outside the circle at any time during their 200 step walk (86 in one particular run of this notebook) and the number that end up outside the circle (392). To get the former number, I used the function "Map" to apply the function "Max" to the list of ant radial positions, and then counting the ones whose maximum radial distances were never larger than "radius". It is simpler to get the former number, only having to apply "Select" to the end points of all the ant paths.

The difference in these two numbers is interesting; it is apparently not uncommon for an ant to wander back into the circle after once stepping outside it. Indeed, this is what we saw when we plotted the path of the first ant.

10.5 CHAPTER SUMMARY

- It is straightforward in MATHEMATICA to generate arbitrary amounts of integer, real, or complex random numbers.

- One popular use for random numbers in Physics is so-called "Monte Carlo" integration or event simulation.
- Often one needs to generate random numbers according to some kind of probability distribution. There are many such distributions built-in to MATHEMATICA.

EXERCISES

10.1 *Follow the example in Section 10.2 to make a table of approximations to* π. *Use logarithmically increasing values of the number of generated points, and include an entry for the percent difference of the approximate value from the correct value. You will also find it useful to refer to the tabulation of approximations to e from Section 9.1.*

10.2 *A ball is thrown with a nominal initial velocity* $v_0 = 20$ *m/sec at an angle* $\theta_0 = 30°$ *from the horizontal. You can ignore air resistance. Find the distance R at which it hits the ground. Then, for 1000 trials, histogram R (i) when* v_0 *is randomly distributed* ± 1 *m/sec about the nominal, (ii) when* θ_0 *is randomly distributed* $\pm 2.5°$ *about the nominal, and (iii) when both are randomly distributed about their nominal values.*

For (iii) use the same number of trials, and use the first number from each of the v_0 *and* θ_0 *lists for the first trials, the second from each for the second trial, and so on. The simplest way to do this is to use "*Table*" to form a list of replacements lists for* v_0 *and* θ_0.

Calculate the mean and standard deviation for each case. Check to see to what extent the standard deviations for cases (i) and (ii) add "in quadrature" to give the standard deviation for case (iii). That is, the square root of the sums of the squares for (i) and (ii) should agree with that for (iii).

10.3 *Repeat Exercise 10.2 but use Gaussian distributions for the initial velocity and launch angle. Assume the ranges given now correspond to standard deviations of the distributions.*

10.4 *An exponential probability distribution function might be written as*

$$\text{pdf} = Ae^{-\lambda x}$$

Determine the normalization constant A, assuming the distribution applies for $0 \leq x \leq x_0$. *Generate random numbers that follow distribution, and fit the results to confirm that you get back the values you used to generate these numbers.*

10.5 *Throwing dice is a handy way to study the fundamentals of binomial, Poisson, and normal probability distributions. Actually doing the experiment with a handful of dice is particularly instructive, but you can also simulate it* *with* MATHEMATICA. *Of course, you can also do many more dice throws this way!*

Assume you have a handful of five dice. Throw them all at once, and count the number of dice that have a "one" face up. The probability of getting a one on any particular die is of course 1/6*, a relatively small number. So, the probability of getting, say, 3, 4, or 5 ones should be progressively smaller.*

Now repeat throwing the handful of dice and recording the number m *of ones. After a large number of throws, you can make a histogram of* m*, and these should follow some probability distribution. If I call* $p = 1/6$ *the probability for "success" and* $q = 5/6$ *the probability for "failure", then the probability for getting* m *successes is just* $p^m q^{5-m}$ *times the number of combinations that give you* m *ones. That is, the probability distribution function is*

$$\text{pdf} = \binom{5}{m} p^m q^{5-m}$$

This is called the Binomial Distribution because the pdf is just the m*th term of the expansion* $(p+q)^5$*. In* MATHEMATICA*, the function* "Binomial" *calculates the binomial coefficient* $\binom{5}{m}$.

Perform the experiment by throwing handfuls of five dice. (Of course, you can choose a different number, but I stuck with five to avoid any confusion in the discussion above.) To simulate the experiment, just generate a list of random integers between 1 *and* 6*,* "Partition" *the list into groups of five, and then use* "Select" *and* "Length" *to count the number of ones in each group. Histogram this number and compare the result to the Binomial Distribution. The probability of getting five ones is* $(1/6)^5 = 1/7776$ *so you'll want to simulate the experiment if you want to get some counts in the last bin! In this case, it is worthwhile to learn how to set the height specification in* "Histogram" *to a logarithmic scale.*

The Poisson Distribution (10.6) is derived from the Binomial Distribution for n *chances for success, in the limit where* $p \ll 1$ *and* $n \gg 1$ *and the mean* $\mu = np$ *remains fixed. (This is how you prove that the standard deviation* $\sigma = \sqrt{\mu}$ *for the Poisson Distribution.) This should be a decent approximation to the extent that* $1/6 \ll 1$ *and* $5 \gg 1$*. Also consider the Normal Distribution approximation (10.4).*

Make a combined plot of your data points, compared to the Binomial, Poisson, and Normal Distributions, all normalized to the number of throws. Include a legend so you can tell the curves apart.

10.6 *A radioactive source is made from* 10^{12} *atoms of the isotope* ^{137}Cs*, which emits a single gamma ray when it decays, and has a half life of 30 years. A detector with a 3 cm diameter circular face is placed one meter away from the source, and has a 50% efficiency for detecting these gamma rays. The number of detected photons is recorded every minute. Simulate a histogram of the detected photon numbers in a five hour data run. Show that the average and standard deviation of the distribution agree with what you expect.*

10.7 *Two ants move in one dimension. They both start out at the same point, but each takes steps of one unit randomly in either the positive or negative direction. Measure the separation distance between the two ants as a function of the number of steps. Does the result surprise you? Consider giving the second ant a larger step size than the first. How does the result change?*

10.8 *Two ants move in two dimensions. They both start out at the same point, but each takes steps of one unit randomly in each direction. Measure the separation distance between the two ants as a function of the number of steps. Does the result surprise you? Consider giving the second ant a larger step size than the first. How does the result change?*

CHAPTER 11

Animation

CONTENTS

Real time motion is a powerful way to convey the physics of dynamic systems. It is straightforward to create such effects in MATHEMATICA. This chapter focuses on the functions "Manipulate" and "Animate" as ways to make plots and drawings move in real time.[1] We've in fact already encountered "Manipulate" in Chapter 4, in particular on page 52. Furthermore, "Animate" is essentially the same function, just with the default options changed.

Before we get into these functions, however, we'll take a brief detour to discuss "scoping." For most of your calculations with MATHEMATICA, you don't need to be concerned with the internals of how the program works. However, "Manipulate" and "Animate" might confound you if you don't realize the scope in which variables are calculated and stored, so I'll try to clear this up for you before moving on. We'll next revisit "Manipulate" to set plots in motion, and extend this to "Animate". Various useful options will be discussed, as well as the function "ListAnimate".

The real fun starts with the creation of cartoon animations. That is, we'll see how to draw various geometric shapes, and to add color and texture to them. It is straightforward to add these shapes to ordinary plots. Applying "Animate" to these shapes gives you real time visual effects that are not only striking, but can provide excellent insight to the physics.

This chapter concludes with three Physics Examples. One makes use of moving plots, the second adds a simple moving shape to a trajectory to display the motion, and the third draws and animates a pendulum that moves according to the correct nonlinear dynamics.

[1] For a detailed discussion of these and other functions, see the Wolfram Language Guide on *Dynamic Visualization*.

11.1 SCOPING

This is the first chapter where we have to delve into *variable scoping*. We'll illustrate it with examples using "Manipulate".

Here are two different ways you might try using "Manipulate" to do the same thing, namely plot the motion of the sine wave $5\cos(\pi t)\sin(\pi x)$ over time. First, there is the straightforward approach of just using "Plot" for the function inside "Manipulate", that is

```
Manipulate[
 Plot[5 Cos[Pi t] Sin[Pi x], {x, 0, 2},
  PlotRange -> {-5, 5}],
 {t, 0, 4}]
```

Or, instead, you might want to store the wave form in a variable so that you can use it later for something else later on. That is,

```
wave = 5 Cos[Pi t] Sin[Pi x];
Manipulate[
 Plot[wave, {x, 0, 2},
  PlotRange -> {-5, 5}],
 {t, 0, 4}]
```

The first example produces the following output:

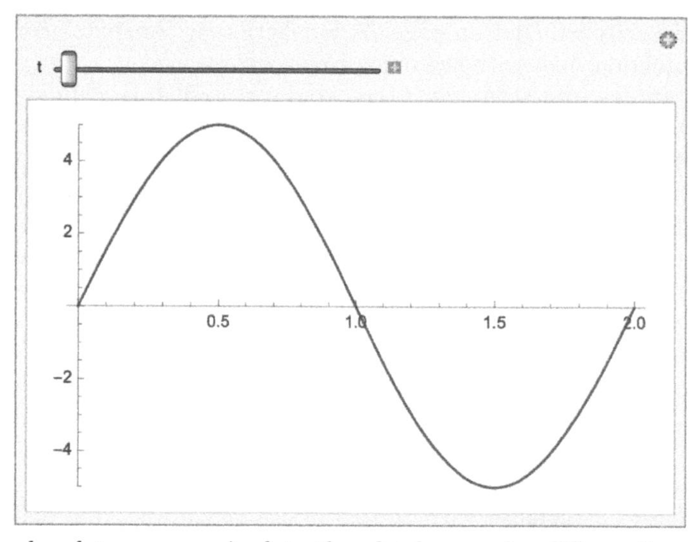

The slider bar lets you manipulate the plot by varying "t" continuously over the range $0 \leq t \leq 4$. This motion is just the first harmonic of a vibrating string, where manipulating t is equivalent to showing the motion at different times.

The second example *does not* give the same output. In fact, it produces an empty plot. The manipulation bar does nothing. Something is wrong.

What's wrong in the second example is that the variable "t" is restricted to the *scope* of the "Plot" command. It is not available to "Manipulate" when it comes time to move the slider bar.

One way to fix this is to use a function to define "wave" instead of using an expression. That is

```
wave[t_] = 5 Cos[Pi t] Sin[Pi x];
Manipulate[
 Plot[wave[t], {x, 0, 2},
  PlotRange -> {-5, 5}],
 {t, 0, 4}]
```

This produces the same output as above, and the manipulation bar works as it should. Defining "wave" as a function creates a variable "t" whose scope extends out to the "Manipulate" function.

Another way to fix the problem is to use the shorthand for "Replace" to replace the variable t with a different variable that extends outside the scope. That is,

```
wave = 5 Cos[Pi t] Sin[Pi x];
Manipulate[
 Plot[wave /. t -> tt, {x, 0, 2},
  PlotRange -> {-5, 5}],
 {tt, 0, 4}]
```

The variable tt is now outside the "Plot" command.

For the purposes of this book, we don't need to get into the nitty gritty of scoping. You can learn more from the MATHEMATICA documentation, but for now, just be aware of potential problems and their easy solution when you are using these dynamic visualization functions.

11.2 PLOTS IN MOTION

To animate the plot from the last section, simply replace "Manipulate" with "Animate" in one of the (working) examples. Choosing the example that uses an expression for "wave" and uses a replacement to get a variable in the appropriate scope, we have

```
wave = 5 Cos[Pi t] Sin[Pi x];
Animate[
 Plot[wave /. t -> time, {x, 0, 2},
  PlotRange -> {-5, 5}],
 {time, 0, 4}]
```

which produces an output that looks like

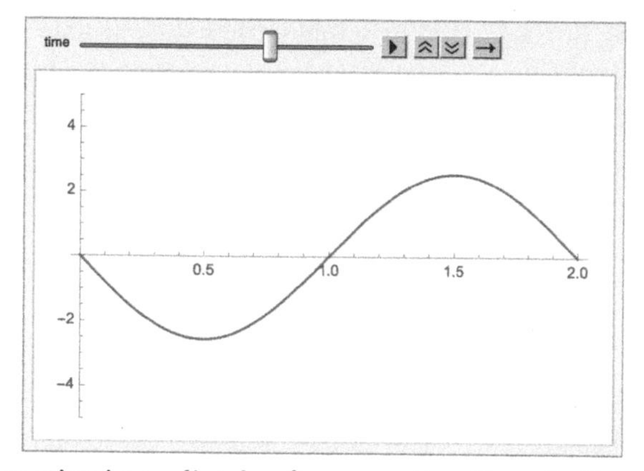

which is set in motion immediately after executing the cell. Stop the animation using the *pause* button to the right of the slider bar. In this image, the *pause* button is replaced by a *play* button. Press this and the animation restarts.

There are many options to go with "Animate" and you should investigate these. For example, setting "AnimationRunning → False" prevents the animation from starting up automatically. You can also change the speed at which "Animation" runs using "AnimationRate".

An associated function is "ListAnimate" which draws a given list of expressions frame by frame. Sometimes you can do the same thing with "Animate" if each frame is derived from the same expression, and you specify discrete steps rather than one continuous animation.

You can also animate plots using "Manipulate". If you replace "Animate" with "Manipulate" in the above example, you get the same output as in the last section. However, if you press the small "+" button just to the right of the slider bar, the image expands into the following:

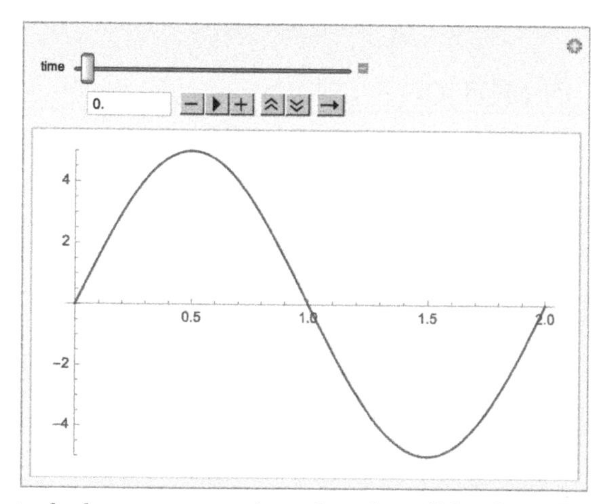

Some new controls have appeared under the slider bar. Pressing the *play* button animates the plot. The box displaying a floating point number gives the value of the manipulated variable.

11.3 DRAWING GEOMETRIC SHAPES

It is possible in MATHEMATICA to create "cartoons" that show how physical systems behave as a function of time. Of course, the first step is to learn how to draw and graph shapes, so let's do that now.

The function you need to learn is "Graphics" which draws graphic primitive objects such as lines, ellipses, and polygons. I'm going to focus on these three here, but you can find a listing (including a host of options) about graphic primitives in the Wolfram Language Tutorial on *Graphics Objects.*

Lines are drawn using the primitive "Line", ellipses with "Circle", and polygons with "Polygon". By themselves, these functions do not create useful output, but as input to "Graphics" they generate the objects you'd expect.

To draw a line from $(x, y) = (-1, -0.5)$ to $(1, 0.5)$, execute the command

```
Graphics[Line[{{-1, -0.5}, {1, 0.5}}]]
```

To draw a circle at the origin and with unit radius, execute the command

```
Graphics[Circle[{0, 0}, 1]]
```

These commands produce the following output:

The generalizations to ellipses and multi-vertex lines are straightforward.

Lines and circles are not necessarily closed figures (you can use "Circle" with the correct options to draw an arc) but polygons are closed by definition. Therefore "Polygon" generates the actual shape, and not the border, by default, and fills it in with black. The defaults, however can be changed.

For example, the three "Graphics" commands in the cell

```
points = {{-1, 0}, {-0.3, 1}, {0.3, 1}, {1, 0}};
trapz = Polygon[points];
Graphics[trapz]
Graphics[{Orange, trapz}]
Graphics[{EdgeForm[Thick], White, trapz}]
```

draw the three figures

You should play with these options, and try the others that are available.

The coordinates used to draw these figures can be tied to other coordinate axes, and combined with other plots. A simple example is

```
y = Sqrt[x];
Show[
 Plot[y, {x, 0, 10}],
 Graphics[Circle[{4, 2}, 0.25]]]
```

which gives the output

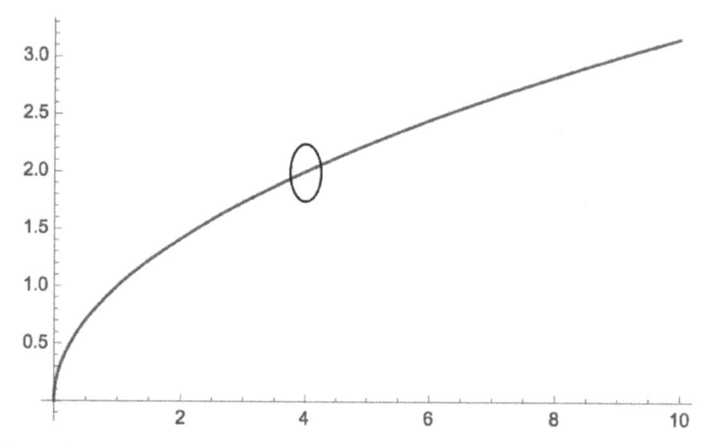

Notice that the circle appears elliptical, because the aspect ratio defined by the "Plot" command is not unity. This can be changed using the "AspectRatio" option in "Plot".

11.4 CARTOON ANIMATION

Now we have what we need in order to create drawings that move. Let's do something simple, and leave the physics examples for the next section.

Consider the following cell:

```
Animate[
 Graphics[
  {Circle[{0, 0}, 4],
   Rotate[Circle[{0, 0}, {1, 4}], angle]}],
 {angle, 0, 10 Pi}, AnimationRunning -> False]
```

Some new things have been added here, so let's go through it piece by piece.

First note that the "Graphics" command is executing a list containing two graphics primitives. The first is simply a circle, centered at the origin, with radius four. The second is a rotated ellipse. We see that "Circle" can be input to a new function "Rotate". In this case, "Circle" is used to draw an ellipse with a long axis for times the length of its short axis. The value of the long axis is chosen so the ellipse fits neatly inside the circle. A variable "angle" is used to specify the rotation angle of the ellipse.

The "Graphics" command is embedded inside "Animate", and "angle" is specified to go through five complete revolutions. We set "AnimateRunning" to "False" so that when the cell executes, the animation does not start automatically.

When you execute this cell, you'll see the following:

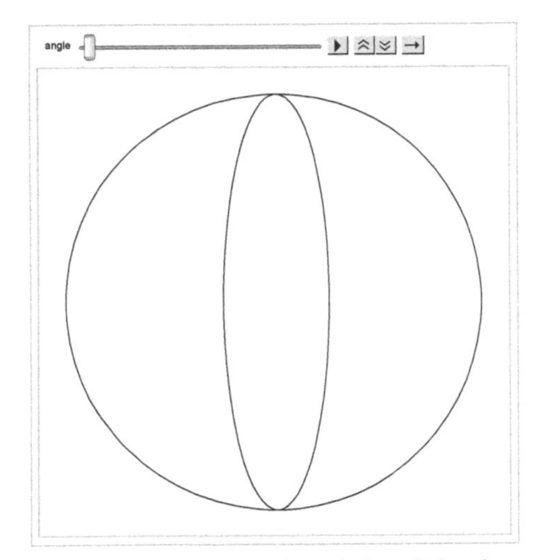

If you press the *play* button to the right of the slider bar, the inner, narrow ellipse will rotate counterclockwise.

As promised, this is a very simple, and useless, animation. However, other than introducing a new function "Rotate" and showing you that you can feed a list of primitives to "Graphics", there is nothing here that you couldn't have done already. Now, however, let's apply this to some physics.

11.5 PHYSICS EXAMPLES

Example 11.1 *Create an animation that shows two finite length wave trains, with the same wavelength, moving towards each other in opposite directions and at the same speed. Choose parameters that make it clear that when the two waves overlap, they generate a standing wave.*

There isn't a lot of physics in this example, but it is worthwhile at least as a nice demonstration that standing waves can be thought of as the interference between two traveling waves.

Notebook 11.1 shows one way to make this animation, and includes the output when the notebook is executed. The first step is to form the rightward– and leftward–moving traveling wave trains, called "trainR" and "trainL", respectively. Clearly the sinusoidal wave form for these are sine or cosine functions of $(x - vt)$ and $(x + vt)$, but we want to truncate these so only a certain length of them are plotted and animated.

Therefore, we multiply the sine waves by the function "UnitBox". As a one-dimensional function, "UnitBox[ξ]" returns unity if $|\xi] \leq \frac{1}{2}$, and zero otherwise. Called with the same $(x \mp vt)$ dependence as its associated sine function, this truncates the sine wave so that it "follows" the wave and the product of the two travels to the right or left.

I've used factors of $\pi/2$ for the sine wave arguments, and divided the arguments of "UnitBox" by "width". Clearly, "width" is just the width of the wave train. I chose $\pi/2$ after some trial and error to come up with a pair of wave trains that had what looked like a suitable number of wavelengths.

Next we set the values of some parameters for the animation. These are the wave velocity "v", the endpoints "end" of the plot, the width of the wave train, and the animation running time, using the parameter "tend". The wave velocity is irrelevant, as the animation will run with whatever speed you set it at, but I included it just to be physically consistent. The wavelength I chose is four units in x, so I chose a range for the plot $(-35 \leq x \leq 35)$ that gives lots of room for the wave trains to move, along with a width that includes five wavelengths in each train. We'll run the animation for a time that corresponds to the endpoints and the speed, reduced a little so that the full wave train fits on the axes at the beginning and end, with a little bit to spare.

Finally, we put this together in the "Animate" command. "PlotRange" is set to not only cover the full range in x, but also to make room vertically so that when the two waves are interfering constructively, they are within range. We override the default and ask that the animation not start running automatically. Notebook 11.1 includes the output. The two wave trains are poised to start moving towards each other, waiting for you to press *play*.

You'll of course need to run this yourself to see how the animation works, but following are snapshots from two different times in the animation, both during overlap of the two wave trains:

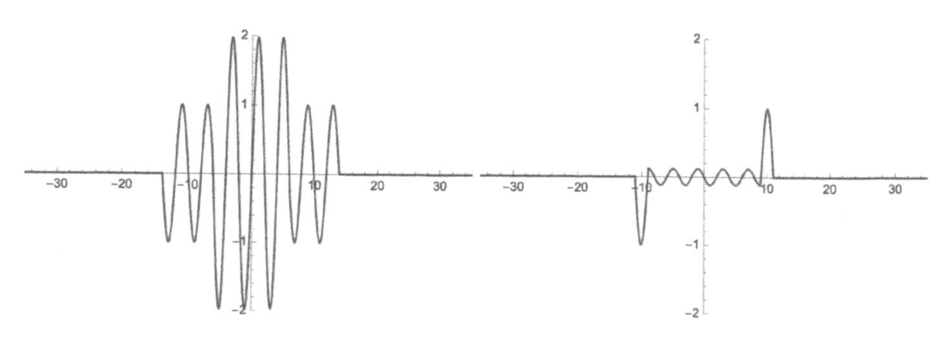

NOTEBOOK 11.1 Solution to Physics Example 11.1.

```
In[1]:= Remove["Global`*"]
```

Standing wave from traveling waves

Form the traveling wave trains

```
In[2]:= trainR = Sin[Pi (x - v t) / 2] UnitBox[(x - v t) / width];
        trainL = Sin[Pi (x + v t) / 2] UnitBox[(x + v t) / width];
        trains = trainR + trainL;
```

Set some parameters

```
In[5]:= v = 1;
        end = 35;
        width = 20;
        tend = (end - 0.6 width) / v;
```

Create the animation

```
In[9]:= Animate[
          Plot[trains /. t → time, {x, -end, end},
           PlotRange → {{-end, end}, {-2, 2}}],
          {time, -tend, tend}, AnimationRunning → False]
```

Out[9]=

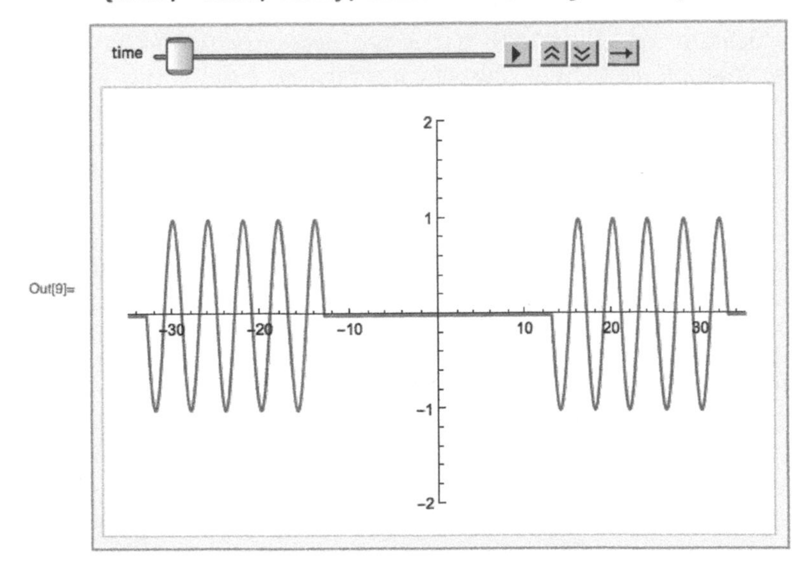

The plot on the left shows the animation at a point where the interference is almost completely constructive, and the right where it is nearly totally destructive. The overlap region indeed oscillates up and down as you expect from a standing wave.

Example 11.2 *An elliptical football is thrown at a 30° angle from the horizontal, with an initial speed of 25 m/sec. Animate the motion of the football, including its tilt along the arc, assuming it was thrown in a "perfect spiral" so that the football's nose is always in the direction of flight. Ignore air resistance.*

The equations that govern the flight of the football are well known from any introductory physics class. We have

$$x(t) = v_0 t \cos\theta \qquad (11.1a)$$

$$y(t) = v_0 t \sin\theta - \frac{1}{2} g t^2 \qquad (11.1b)$$

for the horizontal and vertical positions as a function of time, where v_0 is the initial speed and θ is the initial angle with respect to the horizontal. The velocities v_x and v_y as a function of time are just the derivatives, and the angle $\phi(t) = \tan^{-1}(v_y/v_x)$ is the angle of the football.

My approach to this animation is shown in Notebook 11.2. It starts by writing down Equtions 11.1, deriving the other equations from there. (We could have started by using "DSolve" to solve the simple differential equations.)

The velocity components are obtained from the derivatives, and $\phi(t)$ is calculated. We derive the shape of the path by replacing the time in $y(t)$ with its solution in terms of x. (The result is "path" as a function of "xpath".)

Then we find the maximum height of the football by substituting the time at which the velocity is zero. (It is not necessary to specify the first part of the solution, as there is only one answer, but I did it this way to be consistent with the next line.) The variable "thitsol" is the solution for the time at which $y(t) = 0$, selecting the "Part" that is not $t = 0$. This solution is used to determine the time "thit" at which the football hits the ground, and the "range", that is, the horizontal distance that it travels. Finally, we store the position vector for the football in "point".

Next we put in some numbers, namely the local acceleration due to gravity and the initial speed and angle. I convert degrees to radians the simple way, rather than using "UnitConvert" or other MATHEMATICA commands.

Finally, the animation is produced. Parameters "xlim" and "ylim" are defined to be ten percent larger than the range and maximum height, respectively. These are stored because they will be needed twice in the "Plot" command. Using the "PlotRange" option, they make sure there is some extra space on the right and top to follow the football. We will also use "ylim/xlim" in the "AspectRatio" option so that the plot has the correct dimensions and the path shape is true to real life.

There are two components to the animation, generated by the "Plot" and

NOTEBOOK 11.2 Solution to Physics Example 11.2.

```
In[1]:= Remove["Global`*"]
```

Throwing a football

Define and derive necessary expressions

```
In[2]:= x = v0 Cos[θ] t;
        y = v0 Sin[θ] t - g t^2 / 2;

In[4]:= vx = D[x, t];
        vy = D[y, t];
        ϕ = ArcTan[vx, vy];
        path = y /. Solve[xpath == x, t];
        ymax = y /. Solve[vy == 0, t][[1]];
        thitsol = Solve[y == 0, t][[2]];
        thit = t /. thitsol;
        range = x /. thitsol;
        point = {x, y};
```

Initialize some parameters

```
In[13]:= g = 9.8; v0 = 25; θ = 30 Pi / 180;
```

Produce the animation

```
In[14]:= xlim = 1.1 range; ylim = 1.1 ymax;
         Animate[
          Show[
           Plot[path, {xpath, 0, range}, AspectRatio → ylim / xlim,
            PlotRange → {{0, xlim}, {0, ylim}},
            PlotStyle → Dotted, Axes → {True, False}],
           Graphics[Rotate[
              Circle[point /. t → time, {1.5, 1.0}], ϕ /. t → time]]],
          {time, 0, thit}, AnimationRunning → False]
```

Out[15]=

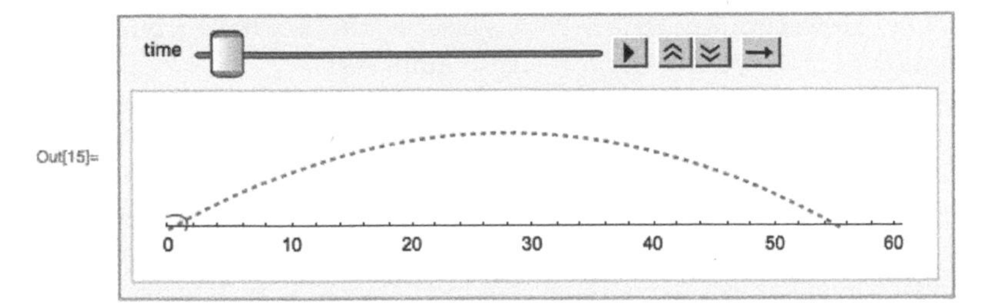

"Graphics" functions inside the "Show" command. I use "Plot" with the appropriate options to set the limits and the aspect ratio. The path is shown with a dotted line, and I turned off the vertical axis. Since the aspect ratio is correct, the vertical scale is the same as the horizontal one.

The "Graphics" command plots the football, at the correct angle, using the "Rotate" and "Circle" functions, just as we did with the example on Page 159. It's a very large football, 1.5 m long and 1.0 m in diameter, but that is only because a realistically sized shape would be very difficult to see at this scale.

Both "point" and "ϕ" are expressions, so to animate the time I use a replacement to get the scope outside of "Graphics" and "Show". The notebook was executed, and includes the output from the "Animate" command. You can just see the nose of the football at the initial point on the path.

Following are two examples of output at different points in the animation:

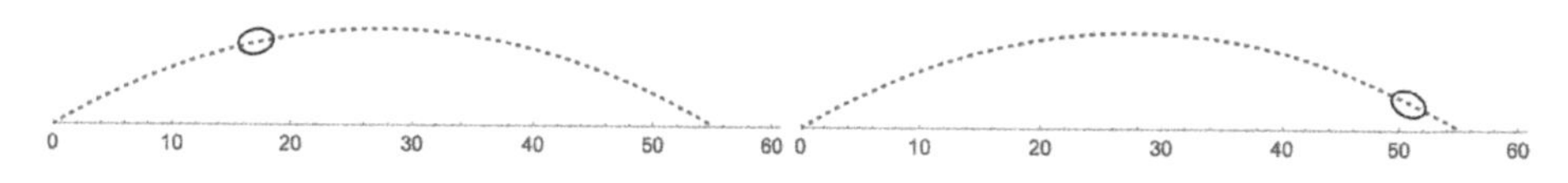

Indeed, the ball follows the path, and tilts at the correct angle so that it is always pointing in the direction of motion.

Example 11.3 *Create a realistic animation of a swinging plane pendulum. Use the correct numerical solution and prove to yourself the motion looks correct when the initial angle is close to the top of the pendulum swing.*

This dramatic and instructive animation is easy to create. Let's first quickly review the physics. A plane pendulum swings through an angle $\phi(t)$ because the torque due to gravity causes an angular acceleration $\ddot{\phi}(t)$. For a mass m at the end of a (massless) rod of length ℓ, the moment of inertia is $\mathcal{I} = m\ell^2$, and the torque is $\tau = -mg\ell\sin\phi$. Therefore $\tau = \mathcal{I}\ddot{\phi}$ becomes

$$\ddot{\phi}(t) + \frac{g}{\ell}\sin\phi = 0 \tag{11.2}$$

Our job is to solve this equation for $\phi(0) = \phi_0$ and $\dot{\phi}(0) = 0$, assuming the pendulum is released from rest.

The "small angle" solution is well known from freshman physics. That is, if $\phi_0 \ll 1$ then $\sin\phi \approx \phi$ and we have simple harmonic motion with period $T = 2\pi\sqrt{\ell/g}$. However, MATHEMATICA gives us the ability to numerically solve Equation 11.2 so we can see a realistic animation.

Notebook 11.3 is one way to create the animation. (I have once again suppressed the output, in order to save space in the figure.) Note that I've written Equation 11.2 in the notebook with $g/\ell = (2\pi)^2$. This means that the small-angle period $T = 1$. This is handy for plotting the solution, which the notebook also does, but irrelevant for the time scale of the animation.

The initial angle is set very large, to just 1% away from the very top of the swing. (Writing it as "ϕ_0=99 Pi/100" is a handy way to play with

NOTEBOOK 11.3 Solution to Physics Example 11.3.

```
In[1]:= Remove["Global`*"]
```

Pendulum animation

Set up and solve the differential equation

```
In[2]:= ϕ0 = 99 Pi / 100;
        tMax = 10;
        eqn = ϕ''[t] + (2 Pi)^2 Sin[ϕ[t]] == 0;
        sol = NDSolve[
            {eqn, ϕ[0] == ϕ0, ϕ'[0] == 0}, ϕ, {t, 0, tMax}];
```

Plot the solution

```
In[6]:= Plot[ϕ[t] /. sol, {t, 0, tMax}, PlotStyle → Thick]
```

Animate the solution

```
In[7]:= len = 1;
        rBob = 0.05 len;
        xBob = len Sin[ϕ[t]] /. sol;
        yBob = -len Cos[ϕ[t]] /. sol;
        pBob = {xBob[[1]], yBob[[1]]};

In[12]:= Animate[Show[
           Plot[0, {x, -len, len}, AspectRatio → 1],
           Graphics[{Thick, Line[{{0, 0}, pBob /. t → time}]}],
           Graphics[{Black, Disk[pBob /. t → time, rBob]}]],
          {time, 0, tMax},
          AnimationRate → 0.5,
          AnimationRunning → False]
```

different starting angles as a fraction of π.) The function “NDSolve” is used to numerically solve the differential equation - Recall Chapter 5 - with the specified initial conditions over ten (small-angle) periods.

Next we plot the solution for $\phi(t)$, a useful check that things are working correctly. For example, you know that if you start a pendulum like this near the top of its swing, the torque will be very small in the beginning and the angle will change only slowly. As the pendulum bob moves further away from vertical, it will speed up and swing through the bottom, slowing down as it gets near the top on the other side of the swing. The period should be much larger than one small-angle period.

Following is the output of the “Plot” command:

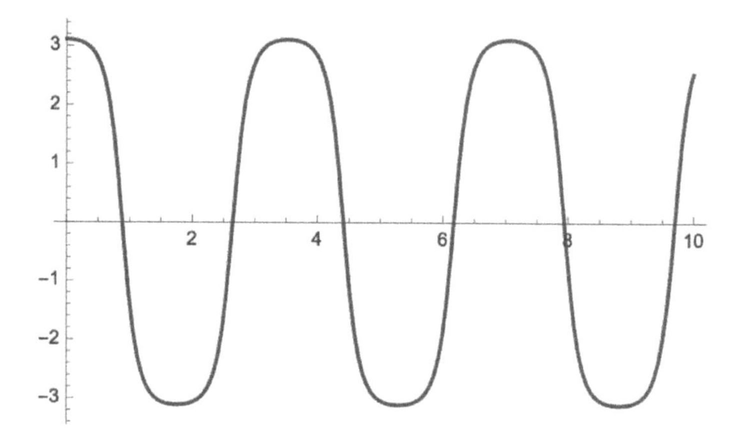

Indeed, this is exactly the expected behavior. Of course, if you start with ϕ_0 small, then the behavior should look like simple harmonic motion.

Now we animate the solution. The result is indeed dramatic. Below I show the initial frame, and two subsequent frames in the same swing:

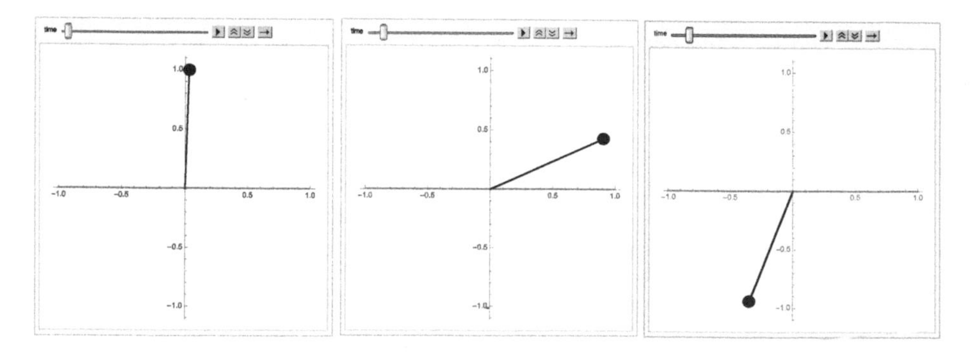

It takes a relatively long time to get from the first frame to the second, but very little time to get from the second frame to the third. You should try adjusting ϕ_0 and watch how the behavior evolves from simple harmonic motion.

Let's go through the animation commands step by step. We define the length of the pendulum rod to be "len=1" and the radius of the bob to be 5% of the length. The x and y positions of the bob are calculated from the solution, and they are stored in a vector "pBob". The "Animate" command is straightforward, using "Show" to combine "Graphics" commands for the rod and the bob with a "Plot" command. I've used "Plot" simply to set the scale and aspect ratio of the pendulum. (I might have done this and included "Axes → False", but I decided to leave the axes on.) I also chose to slow down the animation rate.

11.6 CHAPTER SUMMARY

- It is easy to set plots into motion using "Manipulate" or "Animate", but you need to be careful about the "scope" in which variables are defined.
- You can draw (and customize) standard shapes using graphic primitives. Animating these is also straightforward.

EXERCISES

11.1 *Follow Physics Example 11.1 to make an animation of "colliding opposite triangles." Your animation should start with a graph something like the following:*

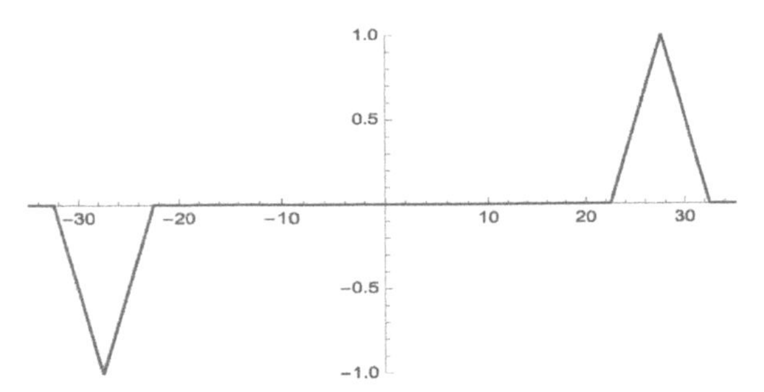

You can use the function "HeavisideLambda" *to make the triangles. Watch what happens when the triangles collide. Is this what you would have expected?*

11.2 *Create an animation of a pendulum where, for a given value of the initial angle $\phi(0)$, $\dot{\phi}(0)$ is chosen so that the pendulum just flips over the top. You can more or less follow Physics Example 11.3, but it will be helpful to write the physical quantities in terms of the length ℓ and gravitational acceleration g when you set up the equations.*

11.3 *Make an animation of a square block, connected to a spring, undergoing simple harmonic motion. The spring is connected to a fixed point at $x = 0$, compressing and expanding with the block. Draw the block with the* "Polygon" *graphic primitive. You can draw the spring as a sine function with some number of wavelengths. You should end up with something like the following, shown at some intermediate time:*

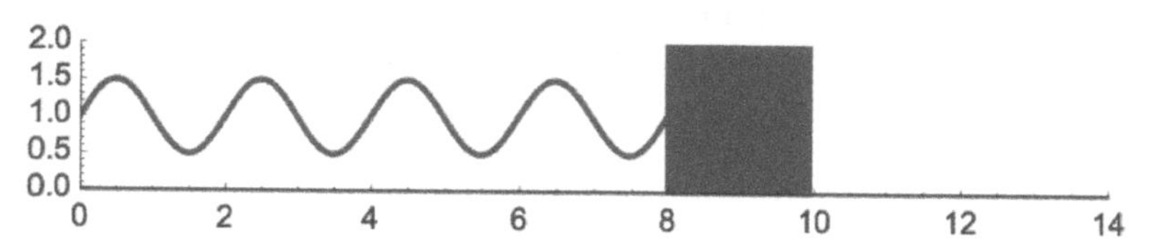

You'll have to make the center of the block, as well as the right endpoint and wavelength of the sine function, depend on time. Use the correct AspectRatio *to make sure the block is drawn as a square.*

11.4 *A one dimensional potential energy well has the form*

$$V(x) = -V_0 \left[e^{-(x-4)^2} + 2e^{-\frac{3}{2}(x-2)^2} \right]$$

giving rise to a force $F = -dV/dx$. A particle starts from rest at $x = 0$. Create an animation of its motion in the well. The particle should be a solid disk that "rides" along the potential energy curve. For example, soon after starting out, the animation should resemble

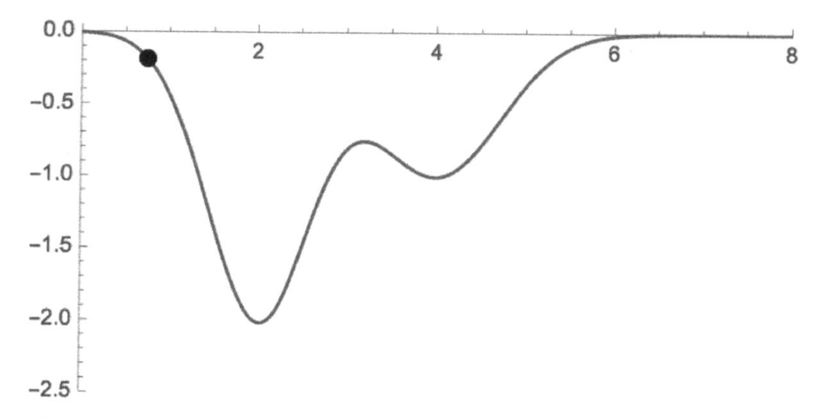

Does the animation behave the way you expect? You might try different initial conditions and see how things change.

11.5 *A resistor $R = 5\ k\Omega$, a capacitor $C = 500\ \mu\mu F$, and an inductor $L = 10\ mH$ are connected in series with a voltage source $V(t) = V_0 \cos \omega t$. At $t = 0$, the charge on the capacitor is zero and there is no current flowing in the circuit. Create an animation that shows a "frequency sweep" of the potential drop $V_R(t)$ across the resistor. The frequency should range logarithmically from $\omega_0/10$ to $10\omega_0$, where $\omega_0 = 1/\sqrt{LC}$. Your plot of $V_R(t)$ should be over $0 \leq t \leq 20\pi/\omega_0$, and fix the vertical scale from $-V_0$ to $+V_0$ throughout the animation. (Choose some value of V_0.) The resonance of the circuit near $\omega = \omega_0$ should be apparent.*

CHAPTER 12

Advanced Plotting and Visualization

CONTENTS

This concluding chapter will show you how to embellish your plots and visualize the results of your calculations using some of MATHEMATICA's most advanced tools. I will also take the opportunity to demonstrate how to use MATHEMATICA to carry out calculations in vector differential calculus. This very useful field of mathematics is generally not covered in the first year physics curriculum, but MATHEMATICA is a good way to learn it and to appreciate its significance for physics.

The first thing we'll do is explore some more of the many options for one-dimensional and two-dimensional plots. We saw some of these options when we introduced "Plot" and "Plot3D", but there are many more. I'll show you some of these, and then let you go off and explore more of them on your own.

Three dimensional surfaces can also be plotted in other ways, such as contour plots and density plots. This chapter will also cover the MATHEMATICA functions that give this kind of output.

After going through the basics of vector differential calculus, we will then explore ways to visualize vector fields. Examples of vector fields include electric and magnetic fields, gravitational fields, and fluid flow. These vector fields are frequently describable in terms of scalar fields called *potentials*. The relationship between a vector field and its potential is best expressed using the machinery of vector differential calculus. I will also use this chapter to give you a brief discussion of polar, cylindrical, and spherical coordinates.

12.1 OPTIONS FOR 2D PLOTS

There are several dozen options for enhancing otherwise simple two dimensional plots. Combining any of these with some other MATHEMATICA functions and you can find better ways to make your point. The documentation will take you through much of these, but here are a few examples that I think might be useful for typical physics applications.

The "Filling" option in MATHEMATICA to set apart different curves on the same set of axes. For example, directly out of the MATHEMATICA documentation,

```
Plot[
 Evaluate[
  Table[
   BesselJ[n, x],
   {n, 4}]],
 {x, 0, 10}, Filling -> Axis,
 PlotLabel -> Bessel Functions, LabelStyle -> {25, Blue}]
```

This produces the following plot:

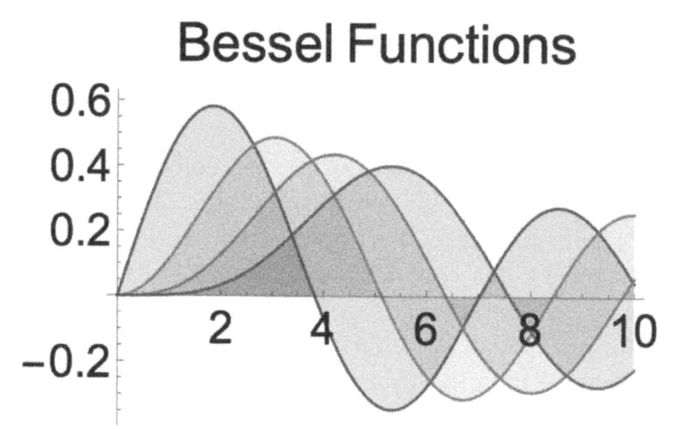

In this case, "Filling" is used to shade in the area between the plotted curve and the axis. You can fill to the top or bottom of the plot, or to some value on the vertical axis. Note that I also decided to make the font size of the tick labels somewhat larger than the default, and to color them blue, and I have included a plot label.

It's worth taking a moment to look at the rest of the cell. Four Bessel functions are plotted, expanding the different orders using "Table". Before handing this over to "Plot", it passes through "Evaluate", which you should read about in the documentation. You should also see what happens if you leave out "Evaluate" and go directly to "Plot".

Sometimes you want to lay out plots side by side, with an alignment of the horizontal axes. You can do this by having plots in a list while using "BaselinePosition" to line up the axes. For example

```
{Plot[Sin[Sqrt[x]] + x/50, {x, 0, 100}, BaselinePosition -> Axis],
 Plot[Sin[Sqrt[x]] - x/50, {x, 0, 100}, BaselinePosition -> Axis]}
```

produces

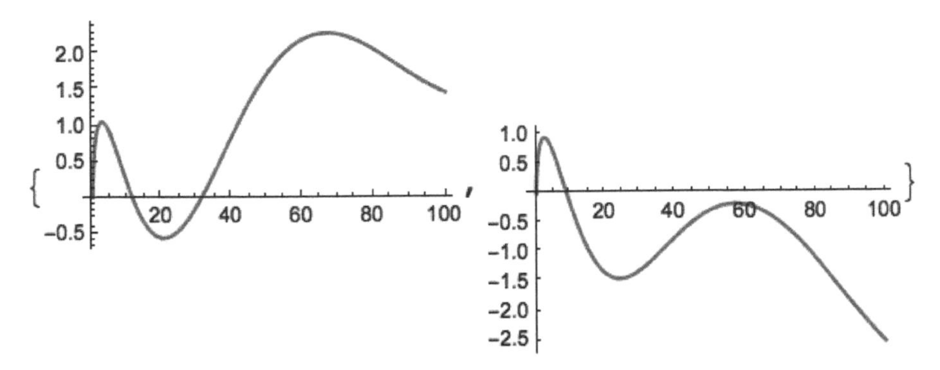

A different way of grouping plots might be to show features of the same information but on different scales, for example linear versus logarithmic. The following example shows how you might demonstrate the tails of two different probability distributions, namely the Gaussian (i.e Normal) and Cauchy[1] distributions, by showing the linear on top of the logarithmic version for each. For example

```
GraphicsGrid[{
  {Plot[PDF[NormalDistribution[0, 1], x], {x, -5, 5},
    Ticks -> None],
   Plot[PDF[CauchyDistribution[0, 1], x], {x, -5, 5},
    Ticks -> None]},
  {LogPlot[PDF[NormalDistribution[0, 1], x], {x, -5, 5},
    Axes -> {True, False}],
   LogPlot[PDF[CauchyDistribution[0, 1], x], {x, -5, 5},
    Axes -> {True, False}]}},
 Frame -> {All, False}]
```

gives the output

[1]Physicists usually refer to the Cauchy distribution as the Lorentz distribution or Lorentzian function.

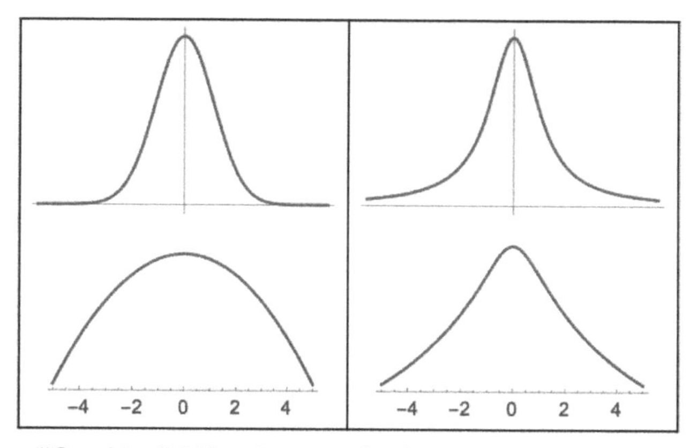

This uses "GraphicsGrid" to lay out the four plots, with an option to frame all of the columns and none of the rows. The plots on the top are linear and I turned the tick marks off for each of these. The logarithmic plots on the bottom, however, retain the full horizontal axes, but do away with the vertical.

Once again, this is just a little taste of what is possible. Explore more with the documentation, and find other examples, to see what else you can do.

12.2 OPTIONS FOR 3D PLOTS

We first introduced three dimensional visualization in Section 5.3, with a very cursory discussion. In this section we will look at what options are available for "Plot3D". We consider contour and density plots in the next section.

For this discussion, let's plot the so-called Higgs potential, which is a model for inducing masses of the elementary particles through the Higgs Mechanism. Here is a cell that defines the function and calls "Plot3D" with all the defaults:

```
\[CapitalPhi] = Sqrt[x^2 + y^2];
v = (\[CapitalPhi]^2 - a^2)^2;
 a = 1; xylim = 1.2 a;
Plot3D[v, {x, -xylim, xylim}, {y, -xylim, xylim}]
```

This gives the following output:

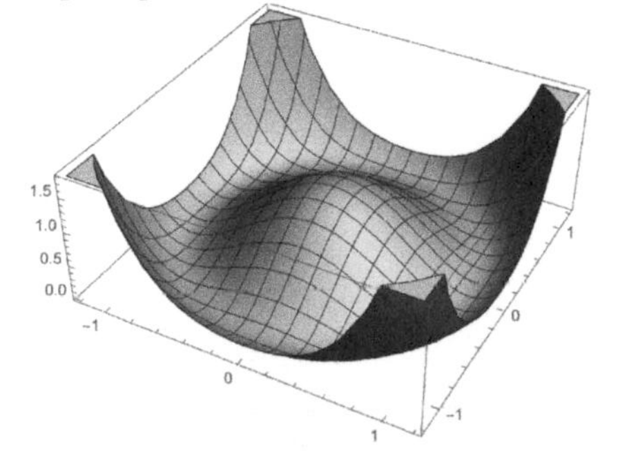

You can specify the point from which the plot is viewed:

```
GraphicsRow[
 {Plot3D[v, {x, -xylim, xylim}, {y, -xylim, xylim},
   ViewPoint -> {0, 2, 2}],
  Plot3D[v, {x, -xylim, xylim}, {y, -xylim, xylim},
   ViewPoint -> Front]}]
```

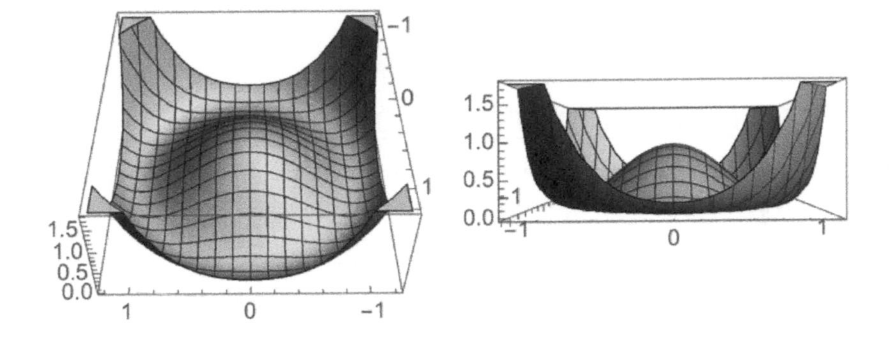

There are many ways to adjust how the three dimensional plot is viewed. These options are described in the documentation on "Graphics3D". An interesting approach you should try is using "Manipulate" to adjust the viewing angle interactively. The following cell is one way to do that:

```
d = 3;
Manipulate[
 Plot3D[v, {x, -xylim, xylim}, {y, -xylim, xylim},
  ViewPoint -> {d Cos[\[Phi]] Sin[\[Theta]],
    d Sin[\[Phi]] Sin[\[Theta]], d Cos[\[Theta]]}],
 {\[Phi], 0, 2 Pi}, {\[Theta], 0, Pi}]
```

Sometimes you may want to compare three dimensional functions, plotted on the same set of axes. The syntax is straightforward, using a list of expressions instead of just one expression, and the MATHEMATICA defaults are usually good enough. In the following, I've added a flat, horizontal surface at $z = 0.25$, modifying the vertical range and specifying my own names for the two curves.

```
Plot3D[{v, 0.25}, {x, -xylim, xylim}, {y, -xylim, xylim},
 PlotRange -> {0, 1}, PlotLegends -> {"Higgs", "Flat"}]
```

The output is shown here. The "Flat" surface is in blue, while the "Higgs" surface is shown in orange:

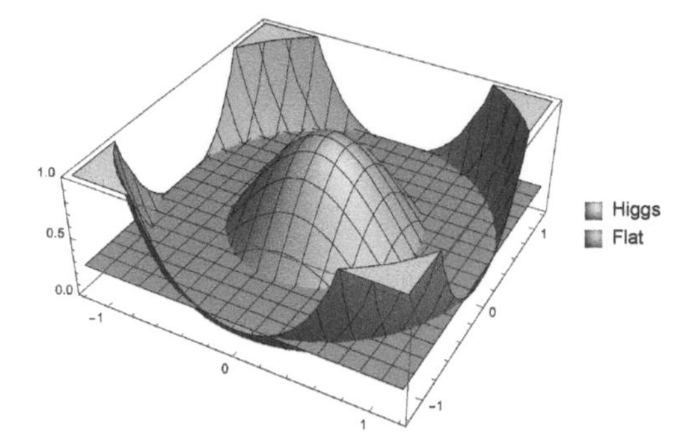

12.3 CONTOUR AND DENSITY PLOTS

Three dimensional renderings of functions of two variables are striking and useful, but frequently you can display more information more effectively using contour plots or density plots. Recall Section 5.3. In this section we will investigate these techniques a bit further.

Let's begin by plotting the "Higgs potential" function from the last section, with the same limits, but using the functions "ContourPlot" and "DensityPlot" instead of "Plot3D". We'll plot them as a list on one line, and use the defaults except that we include a legend for the scale. Execute the cell

```
{ContourPlot[v, {x, -xylim, xylim}, {y, -xylim, xylim},
  PlotLegends -> Automatic],
 DensityPlot[v, {x, -xylim, xylim}, {y, -xylim, xylim},
  PlotLegends -> Automatic]}
```

to find

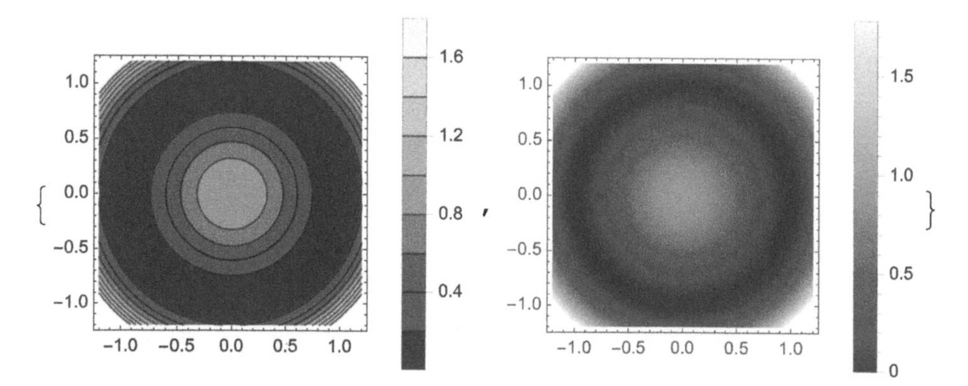

The similarities are obvious. The contour plot has uniform shading between the contours, but the density plot uses a continuous shading scheme, but otherwise, for this case, the information conveyed is essentially the same.

It is possible to specify the contours, and to label them. Executing

```
ContourPlot[v, {x, -xylim, xylim}, {y, -xylim, xylim},
 ContourShading -> None, Contours -> 10]
ContourPlot[v, {x, -xylim, xylim}, {y, -xylim, xylim},
 ContourShading -> None, ContourLabels -> True,
 Contours -> {0.1, 0.5, 0.85}]
```

yields the two plots

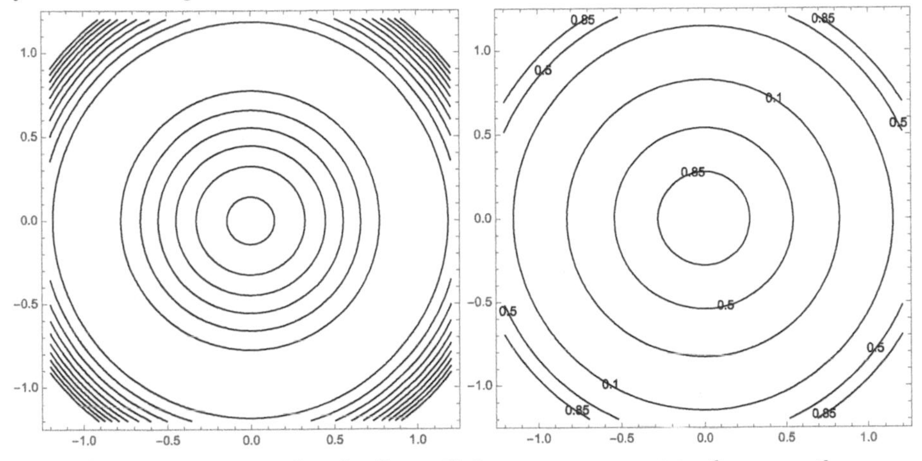

In both cases, we turn the shading off, because we want to focus on the contours. In the first example, we specify ten equally spaced contours on the range of the plot. Since the Higgs potential is steeper as you move away from the center, the contours get closer.

In the second example, we specify the values of the contours we want drawn, and also turn on contour labeling so that we can see the values. (When you run MATHEMATICA interactively, you can move your cursor over the contour and it will reveal the value.)

It is possible to change the font, color, and size of contour labels. See the documentation for some examples, but the idea is to set "ContourLabels" to a pure function that places styled text at some position. For example, the following cell redraws the labeled contour plot above, but with 24pt font size in framed boxes with a white background:

```
ContourPlot[v, {x, -xylim, xylim}, {y, -xylim, xylim},
 ContourShading -> None,
 Contours -> {0.1, 0.5, 0.85},
 ContourLabels -> Function[{x, y, z},
   Text[Style[Framed[z], 24], {x, y}, Background -> White]]]
```

The functions "Text", "Style", and "Framed" have not been introduced before in this textbook, but their use is clear from context. I'm not including the output plot here, but you should try this yourself and play around with the various options.

You can also use "ContourPlot" with a logical argument, to draw contours where the argument returns *True*. For example

```
ContourPlot[v == 0.1, {x, -xylim, xylim}, {y, -xylim, xylim}]
```

returns

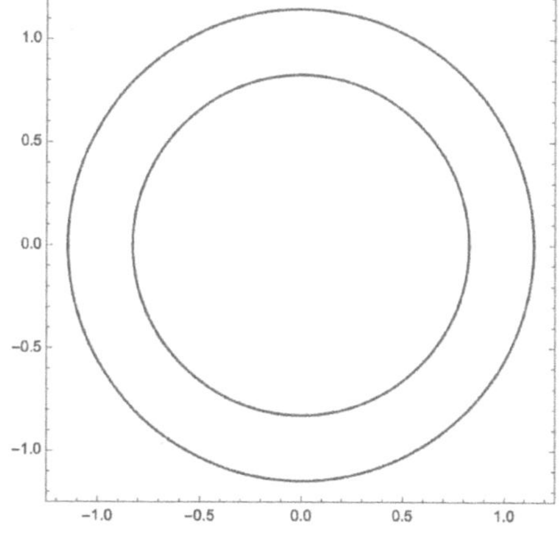

As you expect, there are two circles, clearly on either side of unit radius, where $v = 0.1$. Indeed, "vy0 = v /. y→ 0; Solve[vy0 == 0.1, x]" tells you that the radii of the two circles are 0.83 and 1.15. (If you skipped Chapter 2, you won't understand this, but it is just a solution to a simple algebraic equation.)

Many of the options available to you for "DensityPlot" are similar to those for "ContourPlot", but some are specific. For example, there are a large number of color sets available to you for shading, or you can design your own. It is also possible to display the mesh used to evaluate the shading algorithm. Executing

```
DensityPlot[v, {x, -xylim, xylim}, {y, -xylim, xylim},
 PlotRange -> {0, 1}, ColorFunction -> GrayLevel]
DensityPlot[v, {x, -xylim, xylim}, {y, -xylim, xylim},
 Mesh -> All]
```

produces the two plots

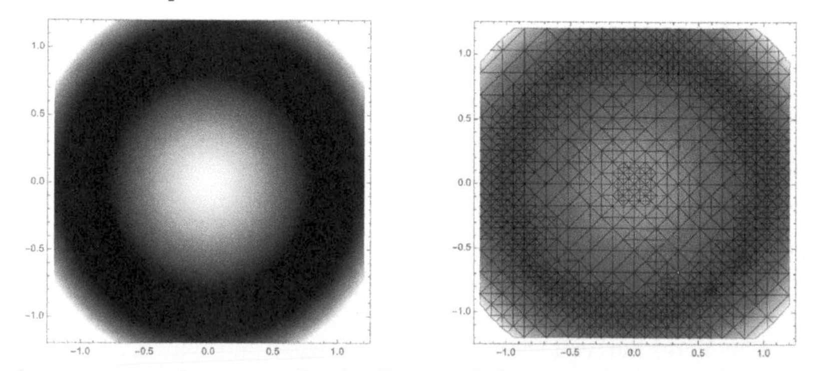

The first uses simple gray scale shading, and the second displays the mesh.

12.4 VECTOR DIFFERENTIAL CALCULUS

This is a good time for another detour for some relevant mathematics, namely working with vector fields in MATHEMATICA. You can reference the Wolfram Language Guide on *Vector Analysis* for more details.

In fact, this mathematics is more advanced than what you'd see in a typical first year physics and math curriculum, but it is essentially in upper level undergraduate material, particularly electromagnetism. If you are familiar with the operators of vector calculus, then most of this section will just be an elementary review that includes references to the relevant MATHEMATICA.

The *gradient* of a function $f(x, y, z)$ is

$$\boldsymbol{\nabla} f \equiv \hat{\mathbf{x}}\frac{\partial f}{\partial x} + \hat{\mathbf{y}}\frac{\partial f}{\partial y} + \hat{\mathbf{z}}\frac{\partial f}{\partial z}$$

where I have assumed the usual three dimensional Cartesian coordinates. The generalization to fewer or more coordinates is obvious, and, as we will discuss soon, one does not need to stick with a Cartesian system.

As will be made clear in Section 12.5, the gradient creates a vector that points in the direction in which $f(x, y, z)$ increases most rapidly.

The MATHEMATICA function that creates a gradient is "Grad" and its application is simple. For example

```
Grad[x y^2, {x, y}]
```

produces the output

```
{y^2, 2 x y}
```

Note that the output is a two-element list (that is, a two-dimensional vector) just as is the second argument to "Grad". (If you have not yet gone through Section 6.1, this would be a good time to do so.) Of course, the generalization of "Grad" to more dimensions is also obvious.

It is useful to think of $\boldsymbol{\nabla}$ as a *vector differential operator*, that is

$$\boldsymbol{\nabla} \equiv \hat{\mathbf{x}}\frac{\partial}{\partial x} + \hat{\mathbf{y}}\frac{\partial}{\partial y} + \hat{\mathbf{z}}\frac{\partial}{\partial z}$$

In other words, think of $\boldsymbol{\nabla} f$ as "operating" on the function $f(x, y, z)$ with the operator $\boldsymbol{\nabla}$. In this sense, we can use $\boldsymbol{\nabla}$ to operate on vector functions as well. If we write

$$\mathbf{v}(x, y, z) = \hat{\mathbf{x}}v_x(x, y, z) + \hat{\mathbf{y}}v_y(x, y, z) + \hat{\mathbf{z}}v_z(x, y, z)$$

then we can define the *divergence* of a vector field $\mathbf{v}(x, y, z)$ as

$$\boldsymbol{\nabla} \cdot \mathbf{v} \equiv \frac{\partial v_x}{\partial x} + \frac{\partial v_y}{\partial y} + \frac{\partial v_z}{\partial z}$$

The *Divergence Theorem*, also known as *Gauss' Theorem*, of vector calculus shows that the divergence is related to the "flow" of the vector field out of some region, hence the name.

You should note in the above notation for the divergence, we made use of the fact that the Cartesian unit vectors $\hat{\mathbf{x}}$, $\hat{\mathbf{y}}$, and $\hat{\mathbf{z}}$ are constants in space and do not depend on the coordinates (x, y, z). Therefore, they are not affected by the partial derivatives. This will not be the case if and when you move to spherical or cylindrical coordinates.

The MATHEMATICA function for divergence is "Div" and the syntax is just what you'd expect. For example

```
Div[{x, y^2, z^3}, {x, y, z}]
```

produces the output

```
1 + 2 y + 3 z^2
```

The other obvious generalization for the operator $\boldsymbol{\nabla}$ on a vector function is called the *curl*, defined as

$$\boldsymbol{\nabla} \times \mathbf{v} \equiv \hat{\mathbf{x}} \left(\frac{\partial v_z}{\partial y} - \frac{\partial v_y}{\partial z} \right) + \hat{\mathbf{y}} \left(\frac{\partial v_x}{\partial z} - \frac{\partial v_z}{\partial x} \right) + \hat{\mathbf{z}} \left(\frac{\partial v_y}{\partial x} - \frac{\partial v_x}{\partial y} \right)$$

In MATHEMATICA, the function "Curl" evaluates the curl of a vector function. Once again, the syntax is just as you'd expect. For example

```
Curl[{z^2, x^2, y^2}, {x, y, z}]
```

returns

```
{2 y, 2 z, 2 x}
```

You can also use "Curl" to get just the z-component of a vector that lies in the xy plane. Again, the syntax is rather obvious. For example

```
Curl[{y^2, x^2}, {x, y}]
```

returns

```
2 x - 2 y
```

The divergence of the gradient is called the *Laplacian*, which is written symbolically as $\boldsymbol{\nabla}^2 \equiv \boldsymbol{\nabla} \cdot \boldsymbol{\nabla}$. So, in MATHEMATICA, if we execute the cell

```
f = Exp[y z] Cos[k x];
Div[Grad[f, {x, y, z}], {x, y, z}] == Laplacian[f, {x, y, z}]
```

we are returned "True". You commonly encounter the Laplacian when solving problems in electrostatics, where the electric potential $u(x, y, z)$ satisfies $\boldsymbol{\nabla}^2 u = 0$ in a charge-free region. For example, with a point charge q at the origin, $u = q/r$ (in the appropriate units) where $r = \sqrt{x^2 + y^2 + z^2}$ is the distance from the charge. In the language of MATHEMATICA, this becomes

```
u = 1/Sqrt[x^2 + y^2 + z^2];
Simplify[Laplacian[u, {x, y, z}]]
```

which returns "0". That is, everywhere (except the origin) is charge free.

If a problem possesses spherical or cylindrical symmetry, it is usually a good idea to work it out in spherical or cylindrical coordinates, and MATHEMATICA has functions that help you with the transformation. Indeed, MATHEMATICA has a large suite of functions for general coordinate transformations, but we will just stick with the basics here. See the Wolfram Language Tutorial on *Changing Coordinate Systems* for more information.

Simple transformations between Cartesian and spherical or cylindrical coordinates[2] are carried out with commands

```
FromSphericalCoordinates[{r, \[Theta], \[Phi]}]
ToSphericalCoordinates[{x, y, z}]
FromPolarCoordinates[{\[Rho], \[Phi]}]
ToPolarCoordinates[{x, y}]
```

The last of these returns

```
{\[Rho] Cos[\[Phi]], \[Rho] Sin[\[Phi]]}
```

while the others also follow just as you would expect. Of course, these transformations are applicable for any vector, not just the position vector.

You can execute vector derivative functions in different coordinates very simply, just by using an optional argument that specifies the coordinate system if it differs from Cartesian. For example,

```
Grad[1/r, {r, \[Theta], \[Phi]}, "Spherical"]
```

returns the vector "{1/r^2,0,0}", and

```
Simplify[Curl[{0, 1/\[Rho]^n}, {\[Rho], \[Phi]}, "Polar"]]
```

returns (equivalently) $(1-n)/\rho^{n+1}$. In the latter example, I specify "Polar" instead of "Cylindrical" because I am using the two-dimensional vector case for "Curl". Also notice that $n=1$ is the case for a two-dimensional magnetic field from a long straight wire on the z-axis, so the curl must be zero.

12.5 VISUALIZING VECTOR FIELDS

To a physicist, a *field* is any function of space (and/or time) with some physical significance. A vector field is one that has both magnitude and direction which can have different values at different positions and different times. Common

[2]We use the physics convention for naming spherical coordinates, namely (r,θ,ϕ) where r is the distance from the origin, θ is the polar angle from the positive z-axis, and ϕ measures the azimuthal angle in the xy plane from the positive x-axis. The cylindrical coordinates use the distance $\rho = r\sin\theta$ in the xy plane along with ϕ and z.

examples of vector fields are fluid velocity profiles, the electric field, magnetic field, and the (Newtonian) gravitational field.

The Wolfram Language Guide on *Vector Visualization* documents the many ways that MATHEMATICA can display such quantities. This section will cover a few basic examples.

There are essentially two approaches. Both approaches plot little arrows on a grid, where the arrows point in the same direction as the field. However, one approach, using the functions "VectorPlot" or "VectorPlot3D", by default, sets the length of the arrows proportional to the magnitude of the field at that point. The second approach uses "StreamPlot" to display the streamlines, or "flow" of the vector field. The information contained in these two approaches is similar, but not equivalent.

Let's start by using "VectorPlot" and "StreamPlot" to each display the same, simple vector field. We can form the field by taking the gradient of some scalar field, which we choose to be $\phi(x, y) = \sqrt{x}$. This field changes only in the x-direction, quickly at first (for $x \geq 0$) and then slows down. The vector field $\mathbf{v}(x, y) = \nabla\phi = (1/2\sqrt{x})\hat{\mathbf{x}}$ embodies these changes. We might call $\phi(x, y)$ an electric potential[3] and $\mathbf{v}(x, y)$ an electric field, albeit with sign changes and disregard for units.

The following cell visualizes this field using "VectorPlot" and "StreamPlot":

```
\[Phi] = Sqrt[x];
v = Grad[\[Phi], {x, y}];
Show[ContourPlot[\[Phi], {x, 1, 10}, {y, -5, 5},
  ContourShading -> None],
 VectorPlot[v, {x, 1, 10}, {y, -5, 5}]]
StreamPlot[v, {x, 1, 10}, {y, -5, 5}]
```

Here are the two output plots:

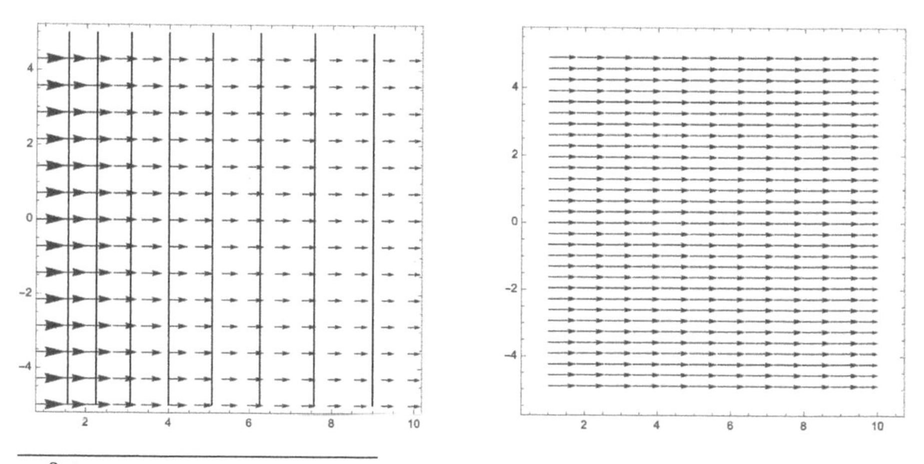

[3]If you haven't yet taken an upper level class in electrostatics, don't be discouraged. In the x-direction, we are just writing that $v = d\phi/dx$, and this will govern the lengths of the arrows in "VectorPlot".

In the first case, we combine a contour plot of the equipotentials with the vector plot. The field values decrease appropriately as the spacing between the equipotentials increases. For the stream plot, all of the arrows are the same length; you can easily see the "flow" of the vector field, but not its strength.

Now we'll try something a little more interesting. Consider the vector field

$$\mathbf{v}(x, y) = (x - 2y)\hat{\mathbf{x}} + (x + y)\hat{\mathbf{y}} \tag{12.1}$$

These components will both increase and decrease, and the directions will shift, as we move away from the origin. We again try visualizing these using both "VectorPlot" and "StreamPlot" as follows:[4]

```
v = {x - 2 y, x + y};
VectorPlot[v, {x, -5, 5}, {y, -5, 5},
 StreamPoints -> 10, StreamStyle -> Red]
StreamPlot[v, {x, -5, 5}, {y, -5, 5}]
```

Executing this cell produces the following plots:

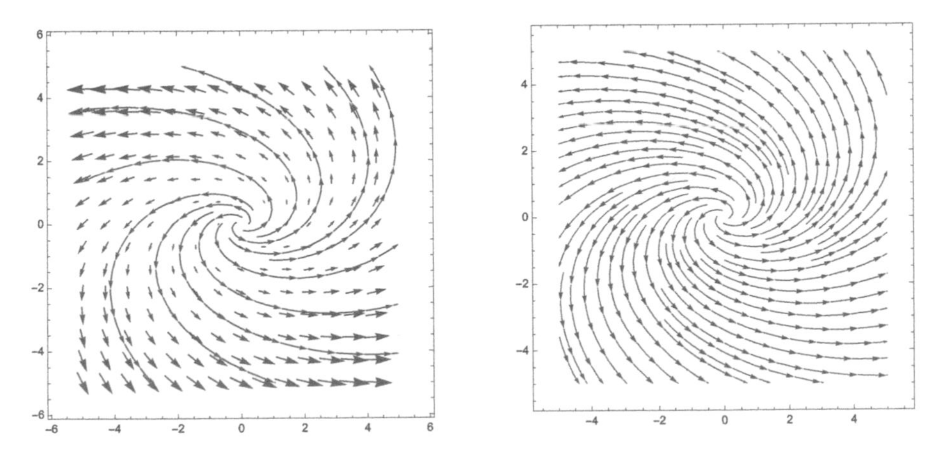

Now the difference between "VectorPlot" and "StreamPlot" is more clear. I have in fact added ten streamlines, in a different color, to the "VectorPlot" output, in order to accentuate the difference.

If the vector field varies by a lot over the viewing region, then you will likely need to override the default options in "VectorPlot" to get a useful visualization. One common example is the "circular" field which decreases in strength in inverse proportion to the distance from the origin. (Introductory physics students would know this as the magnetic field from a long, straight wire carrying a constant current.) Using Cartesian coordinates, we have

$$\mathbf{v}(x, y) = -\frac{y}{x^2 + y^2}\hat{\mathbf{x}} + \frac{x}{x^2 + y^2}\hat{\mathbf{y}} \tag{12.2}$$

[4]We cannot draw "equipotential" lines for this field because $\nabla \times \mathbf{v} \neq 0$.

The field diverges rapidly near $x = y = 0$, so the magnitude of $\mathbf{v}$ varies by large amounts over any plot area that includes the origin. Indeed, the magnitude of this field falls like $1/r = 1/\sqrt{x^2 + y^2}$, as it would if it described the magnetic field from a wire.

Let's once again use "VectorPlot" and "StreamPlot" and compare visualizations. You might try using "VectorPlot" with its defaults on your own, but the output is useless because the divergence at the origin wreaks havoc on the arrow lengths. So, instead, you have to try something like the following:

```
v = {-y/(x^2 + y^2), x/(x^2 + y^2)};
VectorPlot[v, {x, -1, 1}, {y, -1, 1},
 VectorPoints -> 9,
 VectorStyle -> Arrowheads[0.025],
 VectorScale -> {0.125, 1.0, Automatic}]
StreamPlot[v, {x, -1, 1}, {y, -1, 1}]
```

These produce the plots

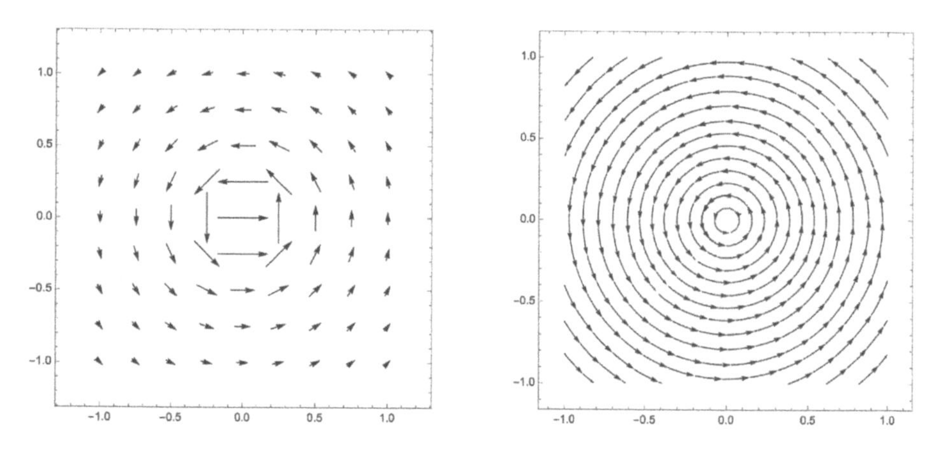

Let's dissect the options in "VectorPlot". We use "VectorPoints" to specify nine points at which to plot arrows, in both the x and y directions. (You can supply a two-element list to specify a different number of points in each direction.) Then we use "VectorStyle" to set the arrowhead size to 0.025 times the width of the graphic; the default is 0.04 and there are many different ways to override this. See the documentation on the graphics directive "Arrowhead" for more information.

We also specify "VectorScale" to specify the arrow sizes. Used in this way, with a three-element list, the meaning is

```
VectorScale -> {length, aspect, scaling}
```

where "length" is the local scale length (that is, the longest arrow length) in terms of a fraction of the plot size; "aspect" is the aspect ratio for the arrowheads (and should perhaps be called "arrowhead ratio" instead); and "scaling" refers to a function that determines the widths of the boxes. For this example, we set the scale length to be 1/8 of the plot size, that is, just a little larger than the 1/9 we used for separation between boxes, a unit aspect ratio, and the default scaling function.

The stream plot is pretty much what you'd expect, equally spaced streamlines in circles about the origin.

We'll stop here, but you should try out other functions described in the *Vector Visualization* guide, in particular "VectorPlot3D".

12.6 PHYSICS EXAMPLES

Example 12.1 *Create a three dimensional plot of the orbits of Jupiter, Saturn, Uranus, Neptune, and Pluto. Include a color scheme with labels.*

This exercise is much simpler than you might think, thanks to the data base access provided in MATHEMATICA. Review Section 9.4 before you try to understand this solution.

My solution is shown in Notebook 12.1. The key to making this simple, is that one of the properties that you can retrieve from the planetary data base, are the actual orbits in three dimensions.

There is one small complication, because Pluto is no longer considered a planet. We use "PlanetData" to retrieve the orbit paths for the major planets, and "MinorPlanetData" to retrieve the orbit path for Pluto. These are flattened into one list, and I also create a flattened list combining the planet names with "Pluto".

The orbit path is literally the graphic primitive "Line" with the orbit data, for each element of the list. It could be fed directly into "Graphics3D", but I wanted to specify the colors. So, I created a list of five colors, and used "Table" set them up with each orbit as a merged list, before feeding this into "Graphics3D". I also made a table of the colors and planets (including Pluto), and use "GraphicsRow" to put this table next to the orbit plot. The output is

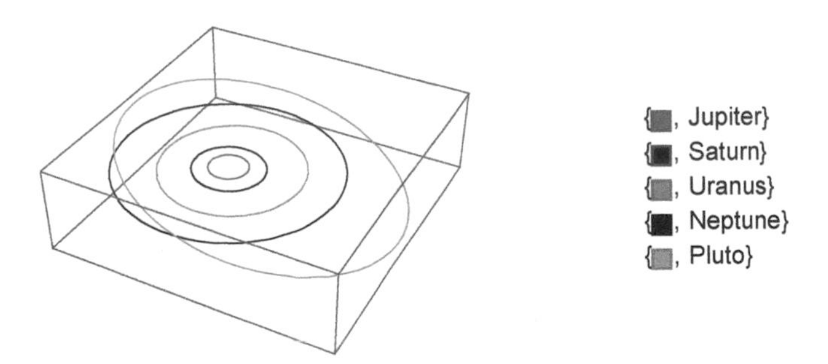

Obviously, it could be embellished further, but this makes the point.

NOTEBOOK 12.1 A solution to Example 12.1.

```
In[1]:= Remove["Global`*"]
```

Orbits of the Outer Planets

Retrieve the planet orbit data

```
In[2]:= planets = {"Jupiter", "Saturn", "Uranus", "Neptune"};
        orbits = Flatten[{PlanetData[planets, "OrbitPath"],
            MinorPlanetData["Pluto", "OrbitPath"]}];
        pluto2 = Flatten[{planets, "Pluto"}];
```

Plot the orbits

```
In[5]:= colors = {Red, Blue, Green, Black, Orange};
        GraphicsRow[{
          Graphics3D[
           Table[{colors[[i]], orbits[[i]]}, {i, 1, 5}]],
          Column[
           Table[{colors[[i]], pluto2[[i]]}, {i, 1, 5}]]}]
```

Example 12.2 *Two electric charges of opposite sign but equal magnitude are separated by some distance. (This is called an "electric dipole.") Create a plot showing the field lines and equipotentials from such system, going out to distances several times the separation of the charges.*

It is easy to write down the equations we need. If I assume that the charges lie on the z-axis, and are separated by a distance d, then we can write the electric potential as

$$V(x, y, z) = \frac{1}{\sqrt{x^2 + y^2 + (z - d/2)^2}} - \frac{1}{\sqrt{x^2 + y^2 + (z + d/2)^2}} \tag{12.3}$$

where the magnitude of the charge, and other overall factors, are irrelevant for the figure we are to create. The electric field is $\mathbf{E} = -\boldsymbol{\nabla} V$.

Notebook 12.2 shows my approach to this exercise. The exercise doesn't state whether we want a two-dimensional or three-dimensional figure, so this notebook creates one version of each.

First we set up the problem and get expressions for the electric potential and electric field. The charges are put at $\pm d/2$ on the z-axis. (We set "d" to a value, so that we can make use of this value later when we plot the positions of the charges.) Next, the position vectors "rp" and "rn" are determined for

NOTEBOOK 12.2 A solution to Example 12.2.

```
In[1]:= Remove["Global`*"]
```

Potential and Field of an Electric Dipole

Calculate for two charges on the z-axis

```
In[2]:= d = 1;
        rp = {x, y, z - d / 2};
        rn = {x, y, z + d / 2};
        v = 1 / Sqrt[rp.rp] - 1 / Sqrt[rn.rn];
        e = -Grad[v, {x, y, z}];
```

Make a two-dimensional plot in the xz plane

```
In[7]:= vv = v /. y → 0;
        ee = Part[e /. y → 0, {1, 3}];
        Show[
         StreamPlot[ee, {x, -3, 3}, {z, -3, 3}],
         ContourPlot[vv, {x, -3, 3}, {z, -3, 3},
          Contours → {-0.45, -0.25, -0.125, 0,
            0.125, 0.25, 0.45},
          ContourShading → None],
         Graphics[{Red, Disk[{0, d / 2}, d / 8],
           White, Text["+", {0, d / 2}]}],
         Graphics[{Blue, Disk[{0, -d / 2}, d / 8],
           White, Text["-", {0, -d / 2}]}]]
```

Make a three-dimensional plot

```
In[10]:= Show[
          VectorPlot3D[e, {x, -3, 3}, {y, -3, 3}, {z, -3, 3},
           VectorPoints → 7,
           VectorScale → {Small, 0.5, 3 &}],
          ContourPlot3D[v, {x, -3, 3}, {y, -3, 3}, {z, -3, 3},
           Contours → {-0.5, -0.1, 0, 0.1, 0.5},
           ContourStyle → Opacity[0.25], Mesh -> None]]
```

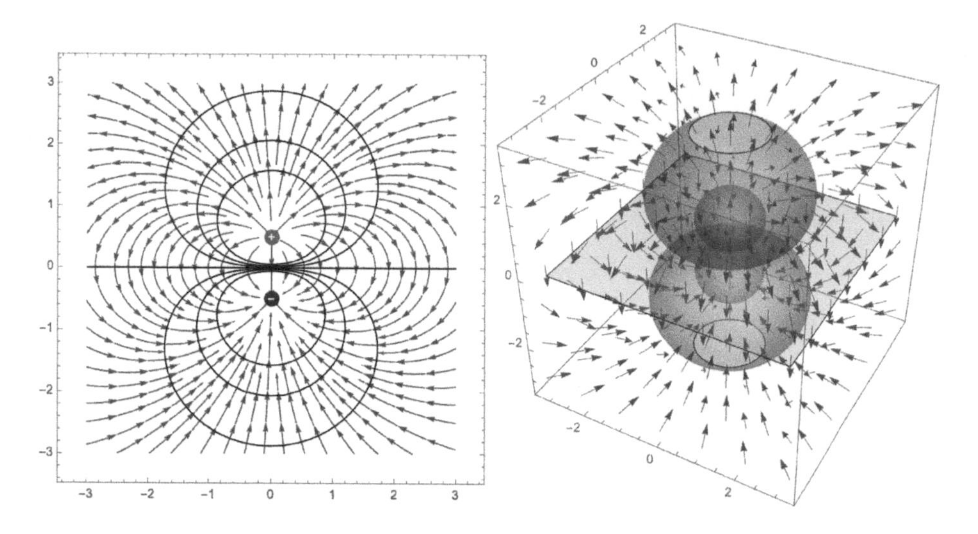

FIGURE 12.1 Output from Notebook 12.2. These are separately the two-dimensional and three-dimensional views of the electric dipole.

any point (x, y, x) in space, relative to the positive and negative charges, respectively. Then, the potential is calculated according to Equation 12.3, using the square root of the inner products of the position vectors in the denominators. The electric field then just follows by taking the (negative of the) gradient of the potential.

The second section of the notebook creates the two-dimensional version. We create two-dimensional versions of the electric potential and electric field (called "vv"and "ee", respectively) by replacing "y" with zero, and taking only the first and third parts (i.e. the x and z components) for the electric field.

Then several images are plotted together with the "Show" function. First, we simply use "StreamPlot" for the field lines, with only the default options. "ContourPlot" plots the equipotentials, and we pick some explicitly for their separation on the figure, with shading turned off. Finally, we use "Graphics" to plot a red disk at the position of the positive charge, and a blue disk at the position of the negative charge. Also in the individual "Graphics" commands, I've placed the graphic primitive "Text" so that white "+" and "−" signs can be inserted in the middle of the disks. The result is shown in Figure 12.1.

For the three-dimensional version, I use "VectorPlot3D" for the electric field. It took some experimentation to come up with vector arrows that I liked. In fact, "Automatic" scaling did not render arrow stems that looked long enough, so I used the simple function "3" to set the arrow length scale. You should experiment with these options yourself.

For the three dimensional contour plot, I again hand picked contour values. I also turned off the mesh on the surfaces, and made sure the surfaces weren't so opaque that I couldn't look through the outer surfaces onto the inner ones.

12.7 CHAPTER SUMMARY

- It is possible to embellish two-dimensional and three-dimensional plots with any number of options to the function. These would be particularly useful for presentations and publications.
- MATHEMATICA provides a set of functions for working with the gradient, divergence, and curl of scalar or vector functions, as well as utilities for changing from Cartesian to spherical or cylindrical coordinate systems.
- Visualizing vector fields with "VectorPlot" and "StreamPlot" is very useful, but these require some care in choosing the options, particularly with rapidly varying vector fields.

EXERCISES

12.1 *Create the following graphic, designed to show the effect of linear drag on the flight trajectory of a baseball, plotted as height versus distance:*

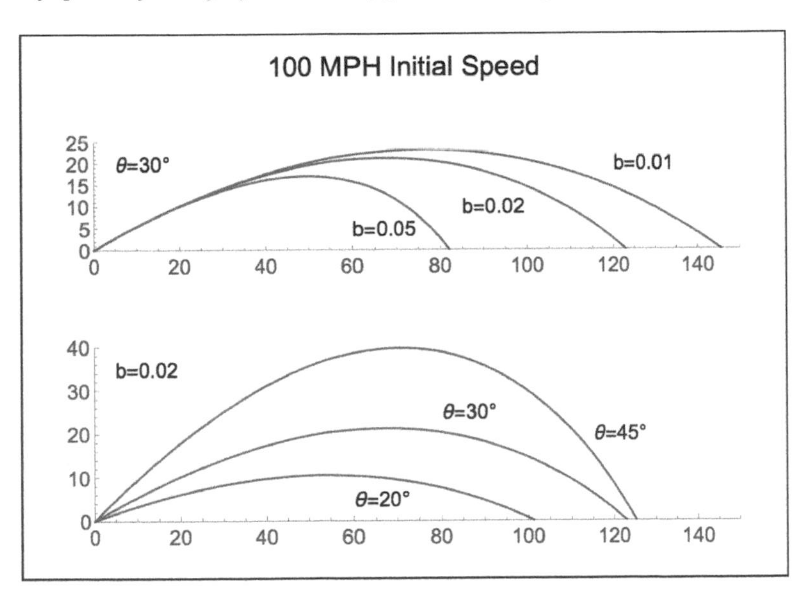

Note that for a drag force $\mathbf{f} = -b\mathbf{v}$*, you can solve the differential equations analytically. The baseball mass is 146 g. All plots are for the same initial speed, but with the indicated values of* b *(in kg/s) and launch angle* θ*.*

*This plot was made using "*Grid*" instead of "*GraphicsGrid*" because it allows control of vertical spacing and we don't need the result to be a graphics object. Three graphics were combined this way, namely the top label and the two plots, and a frame was added.*

*The two plots were made with the same "*PlotRange*" and "*ImageSize*" options so that they line up nicely. Specifically placed text was used to label* *the curves, instead of using legends or some other device to tell the curves apart.*

*The command "*Graphics[Text[Style[. . .]]]*" was used to create the text that is the header of the figure and also the labels in the plot. You should refer to the documentation on how to use this construct to get what you want.*

12.2 *Exercise 6.3 solves for the path of a charged particle in three dimensions under the influence of a uniform magnetic field. Use "*ParametricPlot3D*" to draw the path of the particle to confirm that it is a helix.*

12.3 *With as few lines of code as possible, show that the vector field*

$$\mathbf{E}(z,t) = \mathbf{E_0}\cos(kz - \omega t) \qquad \text{where} \qquad \mathbf{E}_0 = E_{0_x}\hat{\mathbf{x}} + E_{0_y}\hat{\mathbf{y}}$$

simultaneously satisfies the differential equations

$$\nabla^2\mathbf{E} = \frac{1}{c^2}\frac{\partial^2\mathbf{E}}{\partial t^2} \qquad \text{and} \qquad \nabla\cdot\mathbf{E} = 0$$

where $c \equiv \omega/k$. Make use of logical expressions for simplicity and conciseness.

12.4 *Create a two-dimensional image of the electric dipole, as in Notebook 12.2, but which uses "*VectorPlot*" instead of "*StreamPlot*" to display the electric field. Work with the options to make your illustration as clear as possible. In particular, come up with a scaling function that is sensitive to the norm of the vector field, but which allows the vectors to all be displayed over the plot. (The norm is available to you as argument "*#5*" in the "*Vector Scale*" option.)*

12.5 *In the co-rotating frame of two celestial objects, gravity vies with the centrifugal force leading to a number of equilibrium positions, called Lagrange Points, in the orbital plane. For two objects with masses M_1 and M_2, with orbital angular velocity $\boldsymbol{\omega}$, the effective potential at position $\mathbf{r}$ is given by*

$$\Phi(\mathbf{r}) = -\frac{GM_1}{|\mathbf{r}-\mathbf{r}_1|} - \frac{GM_2}{|\mathbf{r}-\mathbf{r}_2|} - \frac{1}{2}|\boldsymbol{\omega}\times\mathbf{r}|^2$$

where $\mathbf{r}_1$ and $\mathbf{r}_2$ are the (fixed) positions of the two objects. For two stars, with $M_1 = M_{\rm Sun}$ and $M_2 = M_{\rm Sun}/4$, undergoing circular orbits, make an appropriate contour plot of Φ and find all the Lagrange points. Which are stable and which are unstable points of equilibrium?

*You will find it helpful to scale distances by the separation a between the stars, and to scale Φ by GM_2/a or something similar. Let the xy plane be the orbital plane, and make some plots of Φ versus x for fixed y and versus y for fixed x, to find a good starting point for defining contours. You might also try using "*Manipulate*" to find the contours that show the Lagrange points.*

Note that the force on a test mass m is $-m\nabla\Phi$, so it is not hard to solve for explicit x and y values for the Lagrange points. This lets you check the positions indicated in your contour plot.

APPENDIX A

Additional Exercises

CONTENTS

This appendix gives some exercises that are not particularly attached to the material in any specific chapter. Of course, you can pick up just about any Physics textbook and work the chapter problems with MATHEMATICA, but here are some that I particularly like, partly because they are more challenging than those earlier in this book. For the most part, these problems combine material from different chapters.

A.1 *The Hoover Dam holds back the water on the Colorado River. The dam is 221 m high and 379 m long. Assuming the dam is a flat vertical wall, find the total force exerted on it by the river. The hydrostatic pressure at some depth d is just $\rho g d$ where ρ is the density of water and g is the acceleration due to gravity. Express the force in equivalent metric tons of weight.*

A.2 *Follow Physics Example 6.2 to find the normal modes of the coupled oscillator in Exercise 4.5. Plot the combinations of $x_1(t)$ and $x_2(t)$ that correspond to these modes, that is, the normal coordinates, for the given initial conditions, and verify that they each have the correct single frequency. Be sure to confirm with a choice of parameters where the masses are not the same.*

A.3 *The equation of motion for a damped forced oscillator is*

$$\ddot{x} + 2\beta\dot{x} + \omega_0^2 x = A\cos\omega t$$

It is convenient to express the motion in terms of the natural period $\tau_0 \equiv 2\pi/\omega_0$ and the driving period $\tau \equiv 2\pi/\omega$. Solve for the motion and plot $x(t)$ including the initial conditions which shows the transient behavior. This is avoided in classes because the math is onerous, but it's easy with MATHEMATICA.

a. *Solve and plot for $0 \leq t \leq 10$, using $\beta = 0.1$, and natural period $\tau_0 = 1$, subject to the initial conditions $x(0) = \dot{x}(0) = 0$. Try first for a driving period $\tau = 2$ and amplitude $A = 1$. Then, try other choices. Consider the cases $\beta < \omega_0$ and $\beta < \omega_0$.*

b. *Use* manipulate *to study the behavior of the forced oscillator. Probably the most dramatic parameter to manipulate is τ, and let it span over τ_0 so that you can observe resonance.*

A.4 *A particle of mass m moves in a Gaussian potential well of the form*

$$U(x) = -U_0 \exp\left[-\frac{x^2}{2\sigma^2}\right]$$

a. *Find the period of oscillation and compare to small oscillations in terms of U_0, m, and σ.*
b. *Pick some values for U_0, m, and σ and use energy conservation to make a plot of the oscillation period as a function of amplitude. You can follow the method in Physics Example 3.2. The amplitude should be expressed in terms of units of σ.*
c. *Solve for and plot the motion $x(t)$ for the mass starting from rest at the values $x(0) = 0.1\sigma, \sigma, 2\sigma,$ and 3σ. Plot the result as $x(t)/x(0)$, and combine the plots on a single graph. Distinguish the different amplitudes by color. Check that the period agrees with what you found in (a) for the lowest amplitude.*

A.5 *The electric field vector $\mathbf{E}(x, y, z, t)$ of an electromagnetic wave traveling in the z-direction has x- and y-components given by*

$$\begin{aligned} E_x &= A\cos(kz - \omega t) \\ E_y &= B\cos(kz - \omega t + \phi) \end{aligned}$$

Such a wave is said to be elliptically polarized.
a. *Create an animation of the vector $\mathbf{E}(x, y, z, t)$ at $z = 0$ rotating over time. Choose A and B to have different values, and set ϕ to something between 0 and $\pi/2$. Include the envelope traced out by the vector. You'll see that the envelope appears to be an ellipse tilted at some angle relative to the x-axis.*
b. *Prove that the envelope is an ellipse. Do this by combining the quantities $u \equiv E_x/A$ and $v \equiv E_y/B$ in such a way that you eliminate dependence on t and end up with a quadratic form of u and v equal to a constant. The uv cross term indicates that the ellipse is tilted.*
c. *Write the quadratic form as the inner product of the row vector (u, v), a symmetric matrix m, and the column vector (u, v). Eliminate the cross term by diagonalizing m. Normalize the matrix of eigenvectors and interpret the result as a rotation matrix, and use this to derive a simple expression for* $\tan 2\theta$, *where θ is the rotation angle. Confirm your answer by rotating the ellipse envelope from Part* **a** *through this angle and showing that the ellipse now is aligned along the x-axis.*

A.6 *An oscillating electric dipole produces electromagnetic radiation. See Introduction to Electrodynamics, 4th Edition, by David J. Griffiths for details. One finds that the radiating electric field at a distance r and time t is given by*

$$\mathbf{E} = -\frac{A}{r}\sin\theta\cos[\omega(t - r/c)]\hat{\boldsymbol{\theta}}$$

where A is a constant and ω is the angular frequency of the oscillating dipole, *which lies in the* $\hat{\mathbf{z}}$ *direction. The angle* θ *is measured from the* $+z$*-axis, and the unit vector* $\hat{\boldsymbol{\theta}} = \cos\theta\cos\phi\;\hat{\mathbf{x}} + \cos\theta\sin\phi\;\hat{\mathbf{y}} - \sin\theta\;\hat{\mathbf{z}}$.

Create an animation showing the time-dependent magnitude of $\mathbf{E}(y, z, t)$ *in the* (y, z) $(\phi = \pi/2)$ *plane, for some choice of* A *and* ω*. (You can just choose* $c = 1$*.) The sequence of frames should resemble the following:*

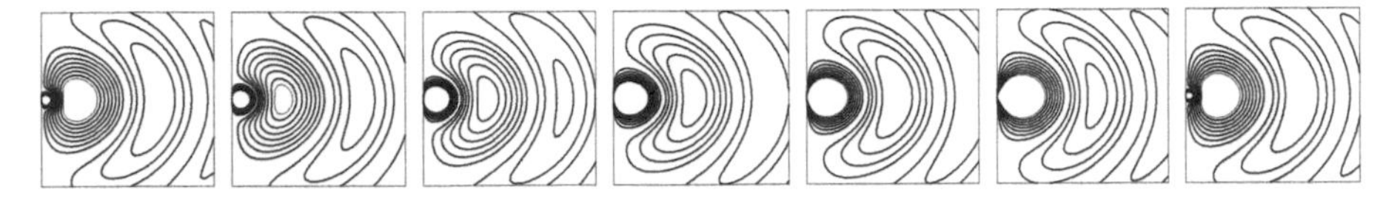

Choose an aspect ratio that agrees with your plot limits in y *and* z*. You may notice that the contours appear "jagged" while the animation is running, but you should find the option which fixes this.*

A.7 *The electrostatic potential at a point* $\mathbf{x}$ *in space from an extended charge distribution can be written as*

$$V(\mathbf{x}) = \left(\frac{1}{4\pi\epsilon_0}\right)\int \frac{dq}{d}$$

where d *is the distance from* $\mathbf{x}$ *to the integration point. The factor out front is for SI units, and is replaced by unity for CGS units. The electric field from that charge distribution is* $\mathbf{E}(\mathbf{x}) = -\boldsymbol{\nabla}V(\mathbf{x})$.

a. *A straight line segment of uniformly distributed charge* Q *and length* L *lies along the* x*-axis, centered on the origin. Find the electrostatic potential along the* z*-axis. Simplify your result as much as you can. Use this to calculate the* z*-component of the electric field along the* z*-axis. Test your results by considering the electric field in the limits* $z \ll L$ *(when it looks as if the line segment is infinitely long) and* $z \gg L$ *(when the line segment appears as a point).*

b. *A flat disk with radius* R *and uniform charge* Q *lies in the* xy *plane, centered on the origin. Repeat as in* **a**.

A.8 *Create the same graphic as in Exercise 12.1, but using quadratic instead of linear drag. The drag force is given by*

$$\mathbf{f} = -\frac{1}{2}\rho v^2 cA\hat{\mathbf{v}} = -\frac{1}{2}\rho v cA\mathbf{v}$$

where $\mathbf{v} = v\hat{\mathbf{v}}$ *is the velocity of the baseball,* $A = \pi r^2$ *is its cross sectional area, and* $\rho \approx 1.2$ *kg/m*^{3} *is the density of air. Reasonable values of the drag coefficient* c*, which should be used to label the curves, are* $c = 0.4,\ 0.5,$ and 0.6. *A regulation baseball is about 75 mm in diameter.*

A.9 *Reconsider Exercise 6.2. Use Equation 6.4 to form the inertia tensor* I*. Look up the documentation for the function "*Outer*" and use this for your calculation. (You'll want to use the "*Times*" function.) Calculate the angular momentum vector* $\mathbf{L}$ *and show that* $\mathbf{L} = \mathsf{I}\cdot\boldsymbol{\omega}$.

A.10 *The files "*CSTarget.dat*" and "*CSBkgrnd.dat*" are two-column lists of data taken with a Compton Scattering experiment with target-in and -out, respectively. The latter ran for twice as long as the former. Read in these files, and confirm the both have the same list in the first column. Collect the bins into groups of five, subtract the background run from the target-in data, and calculate the net error bars on the points. Fit the result to a Gaussian peak plus a linear term, and plot the data and the fit, including error bars. Display the best-fit parameters, the errors on the parameters, determine the χ^2 per degree of freedom, and print out the Analysis-Of-Variance table.*

A.11 *The quantum mechanical wave function of one-electron atoms is given by*

$$\psi_{nlm}(r,\theta,\phi) = N_{nl}e^{-rZ/a_0 n}\left(\frac{rZ}{a_0 n}\right)^l Y_l^m(\theta,\phi)\,{}_1F_1\left(l-n+1;2l+2;\frac{2rZ}{a_0 n}\right)$$

where the normalization constant is

$$N_{nl} = \frac{2^{l+1}}{(2l+1)!}\sqrt{\frac{Z^3(l+n)!}{a_0^3 n^4(n-l-1)!}}$$

and (r,θ,ϕ) are the usual spherical coordinates. The nuclear charge is $+Ze$ and a_0 is a measure of distance called the Bohr radius. The quantum number n can be any positive integer, l is an integer where $0 \le l \le n-1$, and the integer m can take on the values $-l \le m \le +l$. The function $Y_l^m(\theta,\phi)$ is called a spherical harmonic, and the function ${}_1F_1(a;b;z)$ is called the confluent hypergeometric function. Both of these special functions are available in MATHEMATICA.

a. *Write code that forms a general expression for $\psi_{nlm}(r,\theta,\phi)$, making it easy to separate out the normalization factor and the r-dependence. Make use of "*$Assumptions*" so that you can globally declare Z and a_0 to be positive.*

b. *Choose several different values of n, l, and m, and show that each gives a properly normalized wave function, that is gives unit probability over all space as*

$$\int |\psi_{nlm}(r,\theta,\phi)|^2\, dV = 1$$

where $dV = r^2 dr \sin\theta d\theta d\phi$ is the volume element in spherical coordinates. (It is a challenge to demonstrate normalization for the general case, but you can try.)

c. *Plot the normalized radial dependence of $\psi_{nlm}(r,\theta,\phi)$ for $n=1$ and $l=0$, that is ψ_{1s}; $n=2$ with $l=0$ and $l=1$, that is ψ_{2s} and ψ_{2p}; and $n=3$ with $l=0$, 1, and 2, that is ψ_{3s}, ψ_{3p}, and ψ_{3d}. You should be able to check your plots against just about any Modern Physics textbook.*

d. *Create three-dimensional contour plots of the probability density distributions $r^2\sin\theta\,|\psi_{nlm}(r,\theta,\phi)|^2$ for the 1s; $2p, m = +1$; and $2p, m = 0$*

wave functions. Use "ToSphericalCoordinates" to get replacements for Cartesian coordinates. You may encounter some difficulty with the time it takes "ContourPlot3D" to execute. If so, try experimenting "PerformanceGoal" or other options.

A.12 *When a heavy, charged elementary particle passes through matter, it loses energy through collisions with atomic electrons. The kinematics and various atomic effects lead to a very asymmetric energy loss distribution, first calculated by Landau. A well written introduction to energy loss processes and the Landau Distribution can be found in D. H. Wilkinson, Nucl. Instr. Meth. A383(1996)513.*

a. *For an incident particle with charge e and velocity v that traverses a thickness x of matter with atomic number Z, atomic mass A, and density ρ, we define*

$$\xi = \frac{2\pi N_A e^4 \rho}{mv^2}\left(\frac{Z}{A}\right)x$$

where N_A is Avogadro's number, and e and m are the electron charge and mass. Find an otherwise numerical expression for ξ in keV in terms of $\beta = v/c$, Z/A, and the "mass thickness" $S = \rho x$ in mg/cm^2. Note that the expression above for ξ is written in CGS units, and the CGS unit of charge is statcoulombs.

b. *Because of its long tail to high energy loss values, the Landau Distribution is best categorized in terms of the "most probable" energy loss*[1]

$$\Delta_p = \xi\left[\ln\frac{2mc^2\beta^2}{(1-\beta^2)I} + \ln\frac{\xi}{I} + 0.2 - \beta^2\right]$$

where I is the ionization potential of the atomic species. Find the most probable energy deposited in 1.7 mm of silicon by a 10 GeV muon. For I, just use the average of the ionization energies for silicon.

c. *The distribution in energy loss Δ is given by $f(x,\Delta) = (1/\xi)\Phi(\lambda)$ where*

$$\lambda = \frac{\Delta}{\xi} - \ln\frac{2mc^2\beta^2}{(1-\beta^2)I} - \ln\frac{\xi}{I} - 1 + \beta^2 + \gamma_{\mathrm{E}}$$

where γ_{E} is Euler's constant. In MATHEMATICA, *you can get $\Phi(\lambda)$ from the "PDF" of "LandauDistribution[0, Pi/2]". Plot $f(x,\Delta)$ for the energy loss in the case described in* **b**. *Integrate $f(x,\Delta)$ to confirm that it is properly normalized.*

d. *Generate 10^4 values of the energy loss Δ, distributed according to the Landau distribution for the slab of silicon described above. Histogram these values using 1 keV bins, and superimpose a plot of the theoretical distribution $f(x,\Delta)$, appropriately normalized, from* **c**.

[1] There are various versions of this relation. This one is from the 2016 Review of Particle Properties, published online by the Particle Data Group at Lawrence Berkeley National Laboratory. See Equation (33.11) in that reference.

e. *Imagine a set of ten identical silicon slabs, arranged one behind the other, so that the muon passes through each of them. Each slab is configured as a detector which records the energy deposition* Δ. *Generate* 10^4 *events, each of which gives an appropriately distributed energy loss in each detector. From the "truncated mean" of the ten measurements which best approximates the most probable energy loss* Δ_p. *That is, sort each of the ten measurements for an event, and find the mean of the lowest* m *measurements, determining the value of* m *which gives the result closest to* Δ_p.

A.13 *An infinitely long wire lies along the* z*-axis, carrying a current* $i = +100$ *A, generating a magnetic field. For time* $t < 0$*, an electron is accelerated from rest through* 100 *V. At time* $t = 0$ *it is at* $(x, y, z) = (a, 0, -a)$*, with* $a = 1$ *cm, moving in the* $+\hat{\mathbf{z}}$ *direction. Plot* x *versus* z *for* $-a \le z \le 3a$*. Make a three-dimensional plot including the magnetic field vectors, the wire, and the electron's path in three dimensions. (Use different colors for each.) Then, animate the motion of the electron (drawn using "*Sphere*") in three dimensions, also including the magnetic field and the wire. Briefly discuss this result. Repeat for different initial conditions, perhaps starting at the same point but with the initial velocity in the* $+\hat{\mathbf{y}}$ *direction.*

A.14 *This exercise follows Chapter 12 in "Classical Mechanics" by John R. Taylor, University Science Books (2005), and demonstrates chaotic motion in a Driven Damped Pendulum (DDP).*
a. *A pendulum consisting of a mass* m *suspended from a massless string of length* l *swings through an angle* ϕ*. The mass is subject to both a linear drag torque* $-lbv = -l^2 b\dot{\phi}$ *and a driving torque* $lF(t)$ *where* $F(t) = F_0 \cos\omega t$*. Cast the differential equation of motion into the form*

$$\ddot{\phi} + 2\beta\dot{\phi} + \omega_0^2 \sin\phi = \gamma\omega_0^2 \cos\omega t$$

b. *Set up the differential equation with the initial conditions* $\phi(0) = \dot{\phi}(0) = 0$*. Use* $\beta = 0.5\omega_0$ *and* $\omega_0 = 1.5\omega$*, and set* $\omega = 2\pi$ *so that all times will be measured in periods of the driving function.*
c. *Solve and plot the equation of motion for* $\gamma = 0.2$ *over* $0 \le t \le 10$*. (This reproduces Figure 12.2 in Taylor, for* $0 \le t \le 6$*.) Also make a pendulum animation, following Physics Example 11.3, that you can easily adapt for the other solutions below. Show that* $\phi(t)$ *is in fact periodic with period* $2\pi/\omega$ *up to five significant digits for* $t = 4, 5, 6, 7, 8, 9, 10$*.*
d. *Repeat for* $\gamma = 1.06$ *(Taylor Figure 12.4). Solve out to* $t = 40$*, and plot for* $0 \le t \le 15$*. Show five digit periodicity after* $t = 35$*.*
e. *For* $\gamma = 1.073$ *the DDP exhibits "period doubling" (Taylor Figure 12.5) and for* $\gamma = 1.077$ *the one finds three periods (Taylor Figure 12.6). Demonstrate these phenomena with your notebook.*
f. *Now show that for* $\gamma = 1.105$ *there is no evidence of any periodicity (Taylor Figure 12.10). The DDP has become "chaotic."*

g. *A principal test of chaotic motion is the extreme sensitivity to initial conditions. In the linear regime ($\gamma \ll 1$ for the DDP) the difference between two solutions with different starting points should decay away exponentially with the transient solution. However, the behavior for large γ is rather different, and if the motion is chaotic, then the difference actually grows with time, even for tiny changes in the starting point. Plot the difference in solutions for $\phi(t)$ with $\dot{\phi}(0) = 0$ and when $\phi(0)$ differs by (i) 0.1 radians with $\gamma = 0.1$ (the linear regime) and (ii) 10^{-4} for $\gamma = 1.105$ (the aperiodic and presumably chaotic case from* **f** *above). It is best to plot the absolute value of the difference on a logarithmic scale.*

See Taylor Chapter 12 for more calculations you can do with the DDP.

APPENDIX B

Shorthands

CONTENTS

MATHEMATICA provides "shorthands" for many commonly used functions. The following table lists shorthands that are included in this book, and gives simple examples for how they might be used.

Shorthand	Function	Usage example
!	Factorial	"4!" for "Factorial[4]"
×	Cross	"v1×v2" for "Cross[v1,v2]"
.	Dot	"v1.v2" for "Dot[v1,v2]"
/	Map	"Sqrt /@ {a, b, c}" for "Map[Sqrt, {a, b, c}]"
%	Out	"(x + y)^2" then "Expand[%]"
[[ ]]	Part	"list[[i]]" for "Part[list,i]"
/.	Replace (⋆)	"expr/.x→a" for "Replace[expr,x→a]"

(⋆) See documentation on "ReplaceAll"

Quite a number of other shorthands are available, mostly for use with functions. See the Wolfram Language "How To" on Shorthand Notations included with the MATHEMATICA documentations for more information.

Index